T0299122

CAMBRIDGE TRACTS IN MATHEMATICS

General Editors

J. BERTOIN, B. BOLLOBÁS, W. FULTON, B. KRA, I. MOERDIJK, C. PRAEGER, P. SARNAK, B. SIMON, B. TOTARO

CAMBRIDGE TRACTS IN MATHEMATICS

GENERAL EDITORS

J. BERTOIN, B. BOLLOBÁS, W. FULTON, B. KRA, I. MOERDIJK,
C. PRAEGER, P. SARNAK, B. SIMON, B. TOTARO

A complete list of books in the series can be found at www.cambridge.org/mathematics.
Recent titles include the following:

Point-Counting and the Zilber–Pink Conjecture

JONATHAN PILA
University of Oxford

CAMBRIDGE
UNIVERSITY PRESS

CAMBRIDGE
UNIVERSITY PRESS

University Printing House, Cambridge CB2 8BS, United Kingdom

One Liberty Plaza, 20th Floor, New York, NY 10006, USA

477 Williamstown Road, Port Melbourne, VIC 3207, Australia

314–321, 3rd Floor, Plot 3, Splendor Forum, Jasola District Centre,
New Delhi – 110025, India

103 Penang Road, #05-06/07, Visioncrest Commercial, Singapore 238467

Cambridge University Press is part of the University of Cambridge.

It furthers the University's mission by disseminating knowledge in the pursuit of
education, learning, and research at the highest international levels of excellence.

www.cambridge.org
Information on this title: www.cambridge.org/9781009170321
DOI: 10.1017/9781009170314

© Jonathan Pila 2022

First published 2022

A catalogue record for this publication is available from the British Library.

Library of Congress Cataloging-in-Publication Data
Names: Pila, Jonathan, 1962- author.
Title: Point-counting and the Zilber-Pink conjecture / Jonathan Pila.
Description: Cambridge ; New York, NY : Cambridge University Press, 2022. |
Series: Cambridge tracts in mathematics ; 228 | Includes bibliographical
references and indexes.
Identifiers: LCCN 2021060968 (print) | LCCN 2021060969 (ebook) | ISBN
9781009170321 (hardback) | ISBN 9781009170314 (ebook)
Subjects: LCSH: Arithmetical algebraic geometry. | Diophantine equations. |
Modular curves. | Model theory. | BISAC: MATHEMATICS / Number Theory
Classification: LCC QA242.5 .P553 2022 (print) | LCC QA242.5 (ebook) |
DDC 516.3/5–dc23/eng20220215
LC record available at https://lccn.loc.gov/2021060968
LC ebook record available at https://lccn.loc.gov/2021060969

ISBN 978-1-009-17032-1 Hardback

Contents

Preface

This book is based on the Hermann Weyl Lectures given by the author at the Institute for Advanced Study in the autumn of 2018. There are several possible approaches to presenting these topics. The book follows the organization of the lectures, though a lot of detail has been added and further material incorporated.

Part I of this book describes the basic counting result and its application to special point problems, highlighting the various ingredients required to make the strategy work. It concludes with a description of how work on establishing those ingredients has progressed.

Accordingly, a discussion of o-minimality, and a description of the model-theoretic framework in which it fits, is deferred to Part II, so that the line of ideas from point-counting to applications is not obscured. The proof of the Counting Theorem (assuming the r-Parameterization Theorem, which is stated and discussed but not proved here) is followed by a discussion of conjectures and results giving improvements. This part also presents the basic definitions and some key results of the Peterzil–Starchenko development [409, 410, 412] of complex analysis in o-minimal structures.

Part III considers functional transcendence results of Ax–Schanuel type. These are cognate with Schanuel's conjecture. While Ax proved his results by differential algebra (with a second proof by differential geometry), the analogues for Shimura varieties have been proved in the complex settings and use o-minimality and point-counting among other ingredients. Such results are thus both an essential part of the point-counting strategy for diophantine problems and an independent application of point-counting.

The Zilber–Pink conjecture is a recent diophantine conjecture that was arrived at from different motivations, and formulated in different versions, by Zilber [550], Bombieri–Masser–Zannier [90], and Pink [445]. It unifies and sets out a wide-ranging generalization of the classical Mordell–Lang, Manin–Mumford, and André–Oort conjectures. Part IV describes this conjecture, presents a selection of known results, and illustrates further applications of point-counting to such problems. Chapters 16 and 17 address uniformity issues.

Nearly all the results presented here are already in the literature. New elements include some technical improvements to the Counting Theorem (in Chapter 9), a functional version of André's Generalized Period conjecture for elliptic curves (Chapter 17), and a proof that the Zilber–Pink conjecture implies (in an effective way) a uniform version of itself in the basic multiplicative and modular settings (Chapter 24).

Prerequisites and Objectives

This book describes a true circle of ideas going from diophantine geometry to model theory, and then returning, and it addresses a mixture of topics across those areas. Its objective is to provide an accessible entry point to the ideas and techniques. It would not be possible to be self-contained but, as the fundamental ideas are quite elementary, most of the book can be read with just a basic familiarity with the notions of classical algebraic geometry and number theory.

The basic notions of model theory are introduced in the hope of conveying the concerns and themes that animate several of the key ideas, but there are no technical prerequisites. This book is not the place for a systematic introduction to Shimura varieties. Details are concentrated around the simplest examples, while the more sophisticated settings are addressed through a mixture of examples, general description, and references.

The overarching message of this book is the fruitfulness of the interaction between o-minimality and diophantine problems. More concretely, the main ideas we hope to convey are the following:

1. The determinant method at the heart of the Counting Theorem is of a similar nature to the auxiliary functions ubiquitous in diophantine approximation and transcendental number theory [354]. Real variable methods, via o-minimality, give it access to problems that seem inaccessible to complex-variable methods, in particular in higher dimensions.
2. O-minimality is a powerful notion. This book seeks to highlight different facets of its utility.
3. Investigations into the model theory of exponentiation animate the area in several different ways. They are at the source of both o-minimality (real exponential) and the diophantine problems studied (complex exponential).
4. Relatedly, Schanuel's conjecture lies behind several aspects.

5. The point-counting approach is well-adapted to problems of Zilber–Pink type.
6. The success of the point-counting strategy for Zilber–Pink problems hinges on the availability of arithmetic ingredients, such as lower bounds for the size of certain Galois orbits, which in turn often devolve to suitable height upper bounds. Such arithmetic estimates seem to be the deepest obstacle to further progress.

Conventions

We adopt the basic conventions of complex algebraic geometry, as in, for example, [383].

A *projective variety* is identified with its set of $\mathbb{C}$-valued points, and means an (absolutely) irreducible projective complex algebraic set, defined over $\mathbb{C}$. Sometimes it is necessary to consider smaller fields of definition, or weaker irreducibility, and such deviations are indicated.

A *quasi-projective* variety is a Zariski open subset of a projective variety. A *curve* is a quasi-projective variety of dimension one.

A *subvariety* is assumed to be closed in the attendant ambient quasi-projective variety. A subvariety A of B is called a *proper subvariety* of B if the inclusion $A \subset B$ is strict. The notation $A \subset_{\text{cpt}} B \cap C$ means that A is an irreducible component of the intersection of B and C.

As the discussion moves between settings with a complex underlying geometry and those with a real underlying geometry, *dimension* sometimes refers to complex dimension of algebraic or complex analytic varieties, and sometimes to real dimension of sets in $\mathbb{R}^n$. The meaning should always be clear from the context.

We adopt further the basic conventions of analytic number theory, as in, for example, [266].

Big O, little o, $\asymp$, $\ll$, $\gg$, and constants $c(\ldots)$ have indicated dependencies in the main statements, sometimes suppressed within proofs.

Constants need not be the same at each appearance, and are in general ineffective. Effectivity (or its absence) is highlighted at certain points.

The cardinality of a set A is denoted $\#A$, the inclusion $A \subset B$ of sets need not be strict, and $\mathbb{N} = \{0, 1, 2, \ldots\}$ denotes the set of natural numbers.

Acknowledgements

My work in this area began in collaboration with Enrico Bombieri, on questions posed by my former doctoral supervisor Peter Sarnak, while I was a post-doctoral visiting member at the Institute for Advanced Study (IAS) in 1988–9. It developed further during a second visiting membership in 2002–3, and this book is built on the framework of the Hermann Weyl Lectures given there in 2018. I am delighted to acknowledge my triple indebtedness to the IAS for its generous support and hospitality, for the wonderful working environment it provides, and for honouring me with the invitation to deliver those lectures. To Peter and Enrico, I would like to express my deep thanks for their guidance, encouragement, and support over many years.

Interaction and communication are the life-blood of our subject and one of the most enjoyable aspects of our profession. I am grateful to Enrico, along with Philipp Habegger, Jacob Tsimerman, Alex Wilkie, and Umberto Zannier, for our collaborations over a long period on the central subject matter of this book. It is likewise a pleasure to acknowledge the influence of discussion and collaboration with colleagues, including especially Daniel Bertrand, Gal Binyamini, Laura Capuano, Raf Cluckers, Chris Daw, Ziyang Gao, Gareth Jones, David Masser, Ngaiming Mok, Thomas Scanlon, Harry Schmidt, Ananth Shankar, and Boris Zilber.

I thank Alex, Gal, Gareth, Umberto, and Justine Pila (my wife) for their comments on and corrections to the text. Naturally, I take responsibility for all remaining defects. My thanks also to David Tranah at Cambridge University Press for his encouragement of this project and for guiding it through the publication process.

This book is dedicated to Justine, without whom it would not have been written, and to our daughters, Dominique and Davita.

1

Introduction

This book is about certain kinds of diophantine problems. Traditionally, such problems concern solving systems of algebraic equations in specified ways, such as, most classically, in integers or rational numbers. Here, however, the initial focus is on *diophantine problems for non-algebraic sets*. Somewhat surprisingly, there are applications of results about rational points on non-algebraic sets to diophantine problems in the traditional sense, in particular to the *André–Oort conjecture* and related problems falling within the broad *Zilber–Pink conjecture*. Since the applications involve interactions between algebraic and non-algebraic sets, certain questions in *functional transcendence* arise. These questions are related to Ax's results for the exponential function giving functional versions of *Schanuel's conjecture*. The methods used rely on ideas and tools from *o-minimality*, a part of model theory dealing with tame sets in real geometry. These are the main themes of this book.

Geometry Governs Arithmetic

An overarching theme in diophantine geometry is that systems of algebraic equations have few rational solutions unless there is a geometric reason. This is the philosophy of "geometry governs arithmetic" ([280, F.5]) promulgated especially by Lang. Few in this context tends to mean only finitely many. The ur-statement of this idea is the Mordell conjecture [381] of 1922, proved by Faltings [212] in 1983:

A curve of genus at least 2 *has only finitely many rational points.*

An example is any non-singular projective plane quartic curve. A conjecture of Lang [328, I.§§2,3] develops this idea quite precisely for algebraic varieties V of arbitrary dimension in terms of the geometrically defined *special set* $S \subset V$. It predicts that $V - S$ is *Mordellic*; that is, has only finitely many rational points

1

over any finitely generated field, and further that S is not Zariski-dense in V if (and only if) V is geometrically of *general type*. In particular, the rational points are not Zariski-dense in a variety of general type, as was also conjectured by Bombieri in the case of surfaces; see, for example, [280, §F.5.2].

Lang's general conjecture seems far out of reach. Various results show that $V - S$ has few rational points in the weaker sense that, for rational points of V up to a given (sufficiently large) height, those in S outnumber those outside S.

For example, it is expected (see e.g. [272, 1.1.2]) that no positive integer can be written as a sum of two positive fifth powers in two different ways. That is, the positive integer solutions to

$$w^5 + x^5 = y^5 + z^5$$

are all trivial in the sense that $\{w, x\} = \{y, z\}$. This can be viewed as a problem about rational points on the corresponding projective surface. This surface is of general type and the trivial solutions lie on lines that are contained in its special set (see e.g. [279]). Hooley [281] proved that, for positive integers up to a (sufficiently large) size T, there are fewer non-trivial solutions than trivial ones. The result was improved by Browning–Heath-Brown [107], who showed that there are at most

$$\ll_\epsilon T^{13/8+\epsilon}$$

non-trivial solutions in positive integers of size up to T. This is much fewer than the

$$2T^2 + O(T)$$

trivial ones. Such results give some indications towards Lang's general conjecture.

Counting Rational Points on Non-algebraic Sets

Let us now turn to diophantine problems for non-algebraic sets. Consider a transcendental plane curve, say the graph on $[0, 1]$ of a non-algebraic function that is real analytic on an open neighbourhood of $[0, 1]$. Constructions going back to Weierstrass show that such a function may take a rational value at every rational argument. For elaborations of this result, see, for example, [338, 448, 498].

However, it turns out that the heights of the points must increase quite rapidly. Bombieri–Pila [93], motivated by questions of Sarnak [465], used elementary methods to show that various classes of real plane curves have few integer points when counted up to a given height. The same method shows that, on the graph of a real analytic function as above, there are few rational points in a height-density sense (see Theorem 2.3 for the precise sense of "few" here).

This result was later generalized to higher-dimensional non-algebraic sets of certain kinds by the Counting Theorem of Pila–Wilkie [439]. The sets addressed by the Counting Theorem are, more precisely, sets *definable in an o-minimal structure*. The Counting Theorem asserts that such a set Z contains only few rational points that do not lie on some connected positive-dimensional semi-algebraic set contained in Z, and can be viewed as a non-algebraic avatar of "geometry governs arithmetic".

Applications to Diophantine Problems

The Counting Theorem has found applications to diophantine problems in the traditional sense by means of a strategy proposed by Zannier to reprove the Manin–Mumford conjecture (theorem of Raynaud [451, 452]), which was implemented in Pila–Zannier [440]. The same strategy turned out to be efficacious for some Relative Manin–Mumford problems of Masser–Zannier [357], for the André–Oort conjecture, and more generally for instances of the far-reaching Zilber–Pink conjecture, of which all the aforementioned problems are special cases.

The connection between point-counting results for non-algebraic sets and traditional diophantine problems is made via the arithmetic properties of classical special functions.

In the simplest example, the (modified) exponential function

$$e: \mathbb{C} \to \mathbb{C}^\times, \quad e(z) = \exp(2\pi i z)$$

takes rational points in $\mathbb{C}$ to roots of unity in $\mathbb{C}^\times$. Let $F \in \mathbb{C}[X, Y]$. Studying rational points on the complex analytic set

$$\{(z, w) \in \mathbb{C}^2 : F(e(z), e(w)) = 0\}$$

is a possible (though perhaps not too plausible) approach to understanding solutions to $F(x, y) = 0$ in roots of unity (torsion points in the multiplicative group $(\mathbb{C}^\times)^2$). (For a reduction in the opposite direction, from "trigonometric" diophantine problems to ordinary ones, see [157].) Characterizing when there are infinitely many such points is a famous toy problem of Lang [324], introduced in connection with the (then) conjectures of Manin–Mumford and Mordell.

The *modular function* (also known as the j-function), defined and holomorphic on the complex upper half-plane $\mathbb{H} = \{z \in \mathbb{C} : \mathrm{Im}(z) > 0\}$, is given by its *q-expansion*

$$j(z) = \frac{E_4(z)^3}{\Delta(z)} = \frac{1}{q} + 744 + \sum_{n=1}^{\infty} c_n q^n, \quad \text{where } q = e(z),$$

and where $E_4(z)$ (weight 4 *Eisenstein series*) and $\Delta(z)$ (*discriminant function*) are defined by

$$E_4(z) = 1 + 240 \sum_{n=1}^{\infty} \sigma_3(n) q^n, \quad \Delta(z) = q \prod_{n=1}^{\infty} (1 - q^n)^{24},$$

convergent for $|q| < 1$; see [541]. Here $\sigma_k(n)$ denotes the sum of kth powers of the positive divisors of n.

The j-function maps quadratic points in $\mathbb{H}$ (i.e. points $z \in \mathbb{H}$ with $[\mathbb{Q}(z) : \mathbb{Q}] = 2$) to *singular moduli*: the j-invariants of elliptic curves with *complex multiplication* (CM; see Definition 4.5). Singular moduli are algebraic integers and they have rich arithmetic properties studied since the nineteenth century.

Let again $F \in \mathbb{C}[X, Y]$. Studying quadratic points on the complex analytic set

$$\{(z, w) \in \mathbb{H}^2 : F(j(z), j(w)) = 0\}$$

is a possible route to understanding the solutions of $F(x, y) = 0$ in singular moduli. Characterizing when there are infinitely many such points is the basic case of the *André–Oort conjecture*, affirmed in this case by André [6].

Functional Transcendence

In applying the idea indicated above to study, for example, the torsion points on a subvariety $V \subset (\mathbb{C}^\times)^n$, a key role is played by the uniformization

$$e : \mathbb{C}^n \to (\mathbb{C}^\times)^n$$

(we abuse notation by using e to also denote its cartesian powers), and the following question arises. When can $e^{-1}(V)$ contain a positive-dimensional algebraic variety W? This can be equivalently framed as follows:

Suppose $W \subset \mathbb{C}^n$ is an algebraic subvariety. Under what conditions is the image $e(W)$ not Zariski-dense in $(\mathbb{C}^\times)^n$?

A sufficient condition is that $W \subset L$ for some translate of a rational linear subspace: if an equation of the form $\sum a_i z_i = c$ is satisfied by all points of W, where $a_i \in \mathbb{Z}$ are not all zero and $c \in \mathbb{C}$, then $e(V) \subset T$, where T is defined by $\prod x_i^{a_i} = e(c)$. In fact, this condition is necessary, a fact that follows from the results of Ax [20] giving functional versions of Schanuel's conjecture.

In studying the André–Oort conjecture for powers of the modular curve, the corresponding question arises for cartesian powers of the j-function (also denoted by j)

$$j\colon \mathbb{H}^n \to \mathbb{C}^n.$$

This question (and further questions along these lines) arise more generally for Shimura varieties and their uniformization by automorphic functions on Hermitian symmetric domains.

Point-counting provides a method for approaching such questions.

O-Minimality

Point-counting requires that the sets involved are suitably tame. However, it is necessary to work with more general categories than real analytic (or even subanalytic) sets to accommodate the exponential function on horizontal strips in $\mathbb{C}$, or the j-function on its classical fundamental domain.

O-minimality is remarkably well-adapted to these requirements. While sometimes presented as a fulfilment of Grothendieck's vision [249, §5] of a tame topology, o-minimality grew out of work of van den Dries [186] on the model theory of the real exponential function prompted by a question of Tarski [499]. The proof by Wilkie [527] that the real exponential function generates an o-minimal structure affirmed its motivating aspiration to accommodate just such functions. O-minimality affords powerful tools applicable to the corresponding "definable" sets.

PART I

POINT-COUNTING AND DIOPHANTINE APPLICATIONS

2

Point-Counting

The Counting Function

We will count the rational points in a set in real Euclidean space according to their height. Height serves as a convenient complexity measure for rational points.

2.1 Definition   The *height* of a rational number $r = a/b$, where a, b are integers and the fraction is in lowest terms, is

$$H(r) = \max\{|a|, |b|\}.$$

This is extended to tuples $r = (r_1, \ldots, r_n) \in \mathbb{Q}^n$ by setting

$$H(r) = \max\{H(r_i), i = 1, \ldots, n\}.$$

2.2 Definition   For a set $Z \subset \mathbb{R}^n$ and $T \geq 1$, set

$$Z(\mathbb{Q}, T) = \{z \in Z \cap \mathbb{Q}^n : H(z) \leq T\},$$

and then define the *counting function* to be

$$N(Z, T) = \#Z(\mathbb{Q}, T),$$

considered as a function of the height parameter T.

Analytic Curves

The most basic counting result of the type we consider is the following theorem concerning the graph $Z \subset \mathbb{R}^2$ of a function $f : [0, 1] \to \mathbb{R}$ that is real analytic on an open neighbourhood of $[0, 1]$ and *transcendental*, meaning that there is no algebraic relation (over $\mathbb{R}$) satisfied identically by x and $y = f(x)$.

2.3 Theorem ([419, Theorem 9])   *Let Z be as above and $\epsilon > 0$. Then there is a constant $c(f, \epsilon)$ such that $N(Z, T) \leq c(f, \epsilon)T^{\epsilon}$ for all $T \geq 1$.*

9

An estimate of this form for the counting function is our sense of "Z contains few rational points". This result is proved in [419] using the methods introduced in Bombieri–Pila [93], which focussed on counting integer points on the *dilation* of a plane curve, variously assumed to be analytic, sufficiently smooth, or algebraic.

We employ the following mean-value theorem for *alternants* (determinants of the form of $\det\left(\phi_i(x_j)\right)$ for functions $\phi_i, i = 1, \ldots, n$ and points $x_j, j = 1, \ldots, n$).

2.4 Lemma (H. A. Schwarz [478] or [93, p. 342])   *Let D be a positive integer and $J \subset \mathbb{R}$ a compact interval. Let $\phi_1, \ldots, \phi_D \in C^{D-1}(J)$, $x_1, \ldots, x_D \in J$ and set*

$$\Delta = \left(\phi_i(x_j)\right)_{i,j=1,\ldots,D}.$$

Then there exist points $\xi_{ij} \in J$ (intermediate to the points x_i) such that

$$\Delta = V(x_1, \ldots, x_D) \det\left(\phi_i^{(j-1)}(\xi_{ij})\right),$$

where $V(x_1, \ldots, x_D)$ is the Vandermonde determinant.

2.5 Corollary   *With the notation as above we have*

$$|\Delta| \leq c(\phi_1, \ldots, \phi_D) |J|^{D(D-1)/2},$$

where $c(\phi_1, \ldots, \phi_D)$ depends on the maximum sizes of the functions ϕ_i and their first $D - 1$ derivatives on the interval J.

The proof of Theorem 2.3 is a typical transcendence-style argument, involving the "fundamental theorem of transcendence theory" (namely, that there are no integers between 0 and 1) and a "zero estimate".

Proof of Theorem 2.3   Fix a positive integer d and set $D = (d+1)(d+2)/2$. Suppose $\left(x_1, f(x_1)\right), \ldots, \left(x_D, f(x_D)\right) \in Z(\mathbb{Q}, T)$ with $x_1, \ldots, x_D \in J$ some subinterval of I. Apply Corollary 2.5 to these points with the D monomial functions

$$\phi_{ij} = x^i f(x)^j, \quad 0 \leq i, j \leq i + j \leq d$$

to conclude that

$$|\Delta| \leq c(f, D) |J|^{D(D-1)/2},$$

where $c(f, D) = c\left(\phi_{ij}, 0 \leq i, j \leq i + j \leq d\right)$ is the constant in Corollary 2.5.

The denominator of each row of Δ can be cleared by multiplying through by a suitable integer of absolute value at most T^{2d} (the product of the dth powers of the denominators of x_i and $f(x_i)$). Hence (clearing all the rows), if $\Delta \neq 0$, then $|\Delta| \geq T^{-2dD}$.

The crucial point is the **ratio** *of the exponents:* $D(D - 1)/2 \asymp d^4$ *in the estimate for* $|\Delta|$ *and* $2dD \asymp d^3$ *in the height estimate. Thus, if*

$$c(f, D)|J|^{D(D-1)/2}T^{2dD} < 1,$$

then we must have $\Delta = 0$ (the fundamental theorem of transcendence theory), and this holds provided

$$|J| \leq c'(f, d)\, T^{-4dD/D(D-1)},$$

which, after simplification, becomes

$$|J| \leq c'(f, d)T^{-8/(d+3)}.$$

On such an interval J, a determinant of the form of Δ must vanish. This means that if we form the rectangular array $\left(x_k^i f(x_k)^j\right)$, where k indexes rows and $i, j : i+j \leq d$ index columns, using all points $x_k \in J$ for which $\left(x_k, f(x_k)\right) \in Z(\mathbb{Q}, T)$, then this array has rank less than D. Then there exists $a_{ij} \in \mathbb{R}, i+j \leq d$, not all zero, such that (summing over $i + j \leq d$)

$$\sum_{i,j} a_{ij} x_k^i f(x_k)^j = 0, \quad \text{for all } k;$$

that is, the points $\left(x_k, f(x_k)\right) \in Z(\mathbb{Q}, T)$ with $x_k \in J$ all lie on a single real algebraic curve V of degree d determined by the a_{ij}.

Now we consider such intersections $V \cap Z$, where V is a real algebraic curve of degree d. Since f is transcendental, the number of intersection points is finite, and indeed there is a uniform bound $\gamma(f, d)$ on $\#(V \cap Z)$ over all curves V of degree d (this is the zero estimate). This follows, for example, from the fact that Z is definable in an o-minimal structure; see Remark 8.14.2. For Z of this specific (analytic or subanalytic) form, it follows from Gabrielov's Theorem [222] (see alternatively [61]). Since I may be covered by at most

$$c''(f, D)T^{8/(d+3)} + 1 \leq c'''(f, D)T^{8/(d+3)},$$

subintervals J of length at most $c'(f, D)T^{-8/(d+3)}$, we have

$$N(Z, \mathbb{Q}) \leq 2\, c'''(f, D)\, \gamma(f, d)\, T^{8/(d+3)}.$$

The proof is completed by choosing d so that $8/(d + 3) \leq \epsilon$. □

2.6 Remarks

1. For the lemma one does not need analyticity but only $D - 1$ continuous derivatives. So for the theorem one needs somewhat less as well. This may seem a minor point but is important. We require that (i) given $\epsilon > 0$, the graph Z can be divided into finitely many pieces on which it can be parameterized

by functions with sufficiently (but finitely) many bounded derivatives; and
(ii) given d, the cardinalities $\#(Z \cap Y)$, where Y is an algebraic curve of
degree d, are uniformly bounded. O-minimality provides both of these for
its definable sets.

2. A proof of Theorem 2.3 using Siegel's Lemma instead of determinants is
given in [531], also used in [530]. A proof via complex analysis (and Siegel's
Lemma) is given in [351] (see also [353, 354]).

Improved Bounds

Theorem 2.3 cannot be much improved in general, as shown by constructions in
[421, 498] of complex analytic functions for which the ϵ goes to zero arbitrarily
slowly along some (lacunary) sequence of values of T. In the other direction,
it is shown in [498] that, for an entire function, a bound polynomial in $\log T$
always holds on a sequence of T going to infinity (even in arbitrarily long
intervals [234]). A much stronger bound for integer points on curves definable
in $\mathbb{R}_{an}$ is established in [530].

Various results give bounds polynomial in $\log T$ for all T, moreover effective
or even explicit, under additional assumptions on f. See generally [293]. The
main difficulty is to control the growth of the *Bézout bounds* $\gamma(f, d)$ with d. For
complex analytic functions, various conditions ensuring polynomial growth
of Bézout bounds are explored in [156], with corresponding polynomial-in-
$\log T$ improvements of Theorem 2.3. Bézout bounds for solutions of algebraic
differential equations are given in [68].

In the real variable setting, a variant method in [93] enables a bound using
only estimates for the *number* of zeros of successive derivatives of the function
f, rather than the norms of derivatives, and the $\gamma(f, d)$. If one works with
Pfaffian functions (see §8.28), then one has strong bounds [223] on the required
quantities. Bounds polynomial in $\log T$ are obtained in this way in [294, 423].
For real analytic functions, local conditions governing the $\gamma(f, d)$ are explored
in [419], using a mean value theorem [447] for linear homogeneous differential
equations.

Results under other conditions, utilizing transcendence measures of a known
value of f, or suitable growth restrictions, are obtained in [98, 99, 125, 126,
127] (see also [100]). Transcendence measures are exploited in more general
settings in [234]. Results for various classical functions are given in [59, 97,
299, 351], and for certain oscillatory functions in [155]; see also [250].

In the complex setting, the ideas are close to the classical Schneider–
Lang method in transcendence theory, see, for example, [325]. Interpolation

determinants (alternants) were introduced into transcendence theory by Laurent [330]. Independently, the ideas of [419] were used to give proofs of some classical transcendence statements in the real variable setting via determinants in [420]; see also [114; 298, Proposition 5.6; 426].

Complex-variable methods seem to be less flexible for higher-dimensional sets.

Higher-Dimensional Sets

Consider now a set $Z \subset \mathbb{R}^n$. We would like an estimate for the counting function registering that non-algebraic sets have few rational points in our height density sense. There are two points to consider regarding the kind of result one might anticipate.

First, as for plane curves, one cannot hope to do this meaningfully for an arbitrary set, and some tameness assumption is needed along the lines of analyticity. Our condition is that Z is *definable in an o-minimal structure*. We postpone defining this notion until Part II, but we note that it includes sets of a form naturally generalizing the graphs in the planar case.

2.7 Proposition *Let $Z \subset \mathbb{R}^n$ be the union of a finite number of images of maps $\phi : (0,1)^k \to (0,1)^n$, where ϕ is real analytic on an open neighbourhood of $[0,1]^k$. Then Z is definable in an o-minimal structure.*

Proof Such sets are definable in the o-minimal structure $\mathbb{R}_{an}$; see §8.21. □

This gives quite a broad class of sets, though the sets definable in an o-minimal structure are significantly richer, and indeed this is crucial in the applications.

Second, and a new feature when $n \geq 3$, such a set Z of real dimension $k \geq 2$, even if it is non-algebraic, could nevertheless contain semi-algebraic sets of positive dimension, and these might contain many (i.e. not few) rational points. For example, the set

$$Z = \left\{ (x,y,z) \in \mathbb{R}^3 : z = x^y, x, y \in [2,3] \right\}$$

contains, for each rational value of y, a segment of the rational curve $z = x^y$, and every such arc contains $\gg T^\eta$ rational points up to height T for some $\eta = \eta(y) > 0$.

2.8 Definition ([189, p. 1]) A *semi-algebraic set* in $\mathbb{R}^n$ is a finite union of sets each of which is defined by finitely many equations and inequalities between polynomials with real coefficients.

Thus, in the first instance, we count rational points of Z not lying in any connected positive-dimensional semi-algebraic subset of Z, and so make the following definition.

2.9 Definition Let $Z \subset \mathbb{R}^n$. We define the *algebraic part* of Z to be the union of all positive-dimensional connected semi-algebraic subsets $A \subset Z$, and denote it Z^{alg}. The complement $Z - Z^{\mathrm{alg}}$ is referred to as the *transcendental part* of Z, and denoted Z^{trans}.

The connectedness condition is essential. Otherwise, if Z contained a positive-dimensional semi-algebraic set A, then this set, together with any individual point $z \in Z$, would be a semi-algebraic set of positive dimension, and we would get $Z^{\mathrm{alg}} = Z$.

The basic point-counting result is the following. We refer to Theorem 2.10, and its various elaborations such as Theorems 2.13 and 9.14, as the Counting Theorem.

2.10 Theorem ([439, Theorem 1.8]) *Let $Z \subset \mathbb{R}^n$ be definable in an o-minimal structure, and $\epsilon > 0$. Then there is a constant $c(Z, \epsilon)$ such that*

$$N(Z^{\mathrm{trans}}, T) \leq c(Z, \epsilon) T^{\epsilon}$$

for all $T \geq 1$.

2.11 Remarks

1. One can view the theorem as a crude analogue of Lang's general conjecture, with Z^{alg} as a crude analogue of the special set in diophantine geometry. The result says that, away from Z^{alg}, there are few rational points.

2. Since definable sets are more general than the real analytic images we have been using as provisional representatives, even in dimension one, this theorem generalizes Theorem 2.3 to a larger class of real curves.

3. By examining the proof one can do better than simply exclude all of Z^{alg}: only some parts need to be excluded for any given ϵ. Furthermore, the points are contained in few connected semi-algebraic subsets. This is important in some applications.

4. The result is uniform in *definable families*; see Definition 8.6. This is an essential feature required in the proof, but leads to strong uniformities in applications; see Chapter 23.

5. The bound in Theorem 2.10 is qualitative and is what follows from o-minimality. It is natural to ask whether the bound $\ll T^{\epsilon}$ can be improved to $\ll (\log T)^{O(1)}$. This is not possible in general, as already remarked, but

Wilkie conjectured that it should hold in certain special (but central) cases. Various results in this direction are discussed in Chapters 10 and 11.

An estimate of the same form holds for algebraic points up to some given bounded degree $k \geq 1$. This is stated in terms of the *absolute multiplicative Weil height $H(\alpha)$* of an algebraic number that extends the height of a rational number as in Definition 2.1. A definition is given, for example, in [85, Definition 1.5.4]. (So $h(\alpha) = \log H(\alpha)$ is the *absolute logarithmic Weil height*.)

We can then define the counting function for algebraic points of degree $k \geq 1$.

2.12 Definition For a set $Z \subset \mathbb{R}^n$, integer $k \geq 1$, and a $T \geq 1$, we set

$$Z(k,T) = \left\{ z = (z_1, \ldots, z_n) \in Z : [\mathbb{Q}(z_i) : \mathbb{Q}] \leq k, H(z_i) \leq T, i = 1, \ldots, n \right\},$$

$$N(k,Z,T) = \#Z(k,T).$$

2.13 Theorem ([424, Theorem 1.6]) *Let $Z \subset \mathbb{R}^n$ be definable in an o-minimal structure, $k \geq 1$, and $\epsilon > 0$. Then there is a constant $c(Z,k,\epsilon)$ such that*

$$N(k, Z^{\text{trans}}, T) \leq c(Z,k,\epsilon)T^{\epsilon}$$

for all $T \geq 1$.

Theorems 2.10 and 2.13 follow from the stronger version given in Theorem 9.14.

Counting Rational and Integral Points on Algebraic Varieties

The paper [93] also establishes a result on integer points of bounded height on a plane algebraic curve, getting a uniform bound for (irreducible) curves of given degree.

2.14 Theorem ([93, Theorem 5]) *Let $F \in \mathbb{R}[X,Y]$ be irreducible of degree d. Then the number of integer points (x,y) with $F(x,y) = 0$ and $|x|, |y| \leq T$ is at most*

$$c(d,\epsilon)T^{1/d+\epsilon}.$$

The exponent $1/d$ is best possible in view of the example $Y = X^d$. Heath-Brown [271] develops a p-adic version of the method applicable to rational points on projective varieties in all dimensions, in particular giving the analogue of Theorem 2.14 for rational points of height up to T with exponent $2/d + \epsilon$. The exponent $2/d$ is best possible here (same example; non-uniformly such a bound was established in [419]; see also [204]).

The applicability of the *p*-adic methods of [271] to higher-dimensional algebraic varieties suggested investigating the applicability of the real variable determinant approach to higher-dimensional non-algebraic sets. This was instigated in [421], for integer points on the dilation of a subanalytic surface (and then [422] for rational points). The real variable approach is applied to rational points on higher-dimensional algebraic varieties in [347].

These determinant methods (real, *p*-adic, and the global version of Salberger) have been useful in a variety of applications to counting rational points on algebraic varieties (see e.g. [272]), in particular towards the *dimension-growth conjecture* [269, 486]; see [271, Conjectures 1, 2]. This posits an exponent $\dim X + \epsilon$ for the counting function for a projective variety $X \subset \mathbb{P}^n$. In the strongest form one asks for the constant dependent only on n, $\deg X$, and ϵ. This is proven (in the strong form) for $d \geq 4$ in work of Browning, Heath-Brown, and Salberger [108, 464]; see also [123].

The T^ϵ in Theorem 2.14 was improved to a power of $\log T$ in [425] and removed altogether from the corresponding result for rational points in [526]. In higher dimensions, real variable parameterization is used in [77] to replace the ϵ by a power of $\log T$, with polynomial dependence of the constants, for rational points on hypersurfaces. This is further refined in [123], using the global determinant method, eliminating the ϵ factor entirely in the dimension-growth conjecture while retaining polynomial dependence of the constants; see further [403].

W. M. Schmidt [477] conjectured that, for curves of positive genus, one has a bound of the form of Theorem 2.14 but with exponent ϵ instead of $1/d + \epsilon$. Here one has finiteness for an individual curve by a famous theorem of Siegel, and the issue is the uniformity. Some progress towards this for elliptic curves is obtained in [273]. A T^ϵ bound for integral points on moduli spaces of varieties is proved in [206]. One would also expect stronger estimates for rational points on curves of positive genus, and such an improvement is obtained in [207]. Under a suitable hypothesis on ranks of elliptic curves, a uniform bound $N(Z, T) \ll_\epsilon T^\epsilon$ for rational points on non-singular cubic plane curves is established in [270].

Counting Rational and Integral Points on Sufficiently Smooth Curves

The paper [93] also considered functions with a finite number of derivatives, generalizing results of Jarnik [291].

2.15 Theorem ([93, Theorem 7])   *Let $f \in C^\infty([0,1])$ be strictly convex with graph Γ and let $\epsilon > 0$. Then, for $t \geq 1$,*

$$\#(t\Gamma \cap \mathbb{Z}^2) \leq c(f, \epsilon) t^{1/2+\epsilon}.$$

Further conjectures and results in this direction are given in [93] and [419, 477]. See also [288] on estimating integer points on or near a plane curve, and references there for such problems with rational points. On rational points near a definable set, see [260] and Theorem 9.17.

3

Multiplicative Manin–Mumford

Statement

As a prototypical example of the kind of diophantine questions for which the counting strategy can be implemented, and in particular of the kind of set to which the Counting Theorem can be applied, we begin, following [469], with the simplest example of the diophantine problems to be considered later.

The most basic case, already mentioned, concerns the multiplicative group $\mathbb{G}_m^2$, where $\mathbb{G}_m = \mathbb{G}_m(\mathbb{C}) = \mathbb{C}^\times$ is the multiplicative group of non-zero complex numbers, and its subgroup of torsion points

$$\left(\mathbb{G}_m^2\right)_{\text{tor}} = \{(\zeta, \eta) : \zeta, \eta \text{ are roots of unity}\}.$$

3.1 Theorem (Lang [324, p. 230], with proofs due to Ihara, Serre, and Tate)
Let $V \subset \mathbb{G}_m^2$ be a curve, defined by $F(x, y) = 0$ for some $F \in \mathbb{C}[X, Y]$. Then $V \cap \left(\mathbb{G}_m^2\right)_{\text{tor}}$ is a finite set unless F is (up to scaling) of the form $X^n Y^m - \xi$ or $X^n - \xi Y^m$, where $n, m \in \mathbb{N}$, not both zero, and ξ is a root of unity.

Note that $V \cap \left(\mathbb{G}_m^2\right)_{\text{tor}}$ is indeed infinite in the exceptional cases.

The generalization of this theorem to $\mathbb{G}_m^n$ is a very special case of the multiplicative Mordell–Lang conjecture (theorem of Laurent [329]; see conjecture 18.3) but can be deduced from results of Mann [342] (see e.g. [157] or [190, 1.1]). For an independent proof via point-counting, see [465] (unpublished; cf. [466]); see also [476].

To state the generalization to $\mathbb{G}_m^n$, we need to introduce its special subvarieties. A detailed treatment of algebraic tori may be found in [85, Ch. 3]. If $\Lambda \subset \mathbb{Z}^n$ is a lattice, then the equation $x_1^{\lambda_1} \ldots x_n^{\lambda_n} = 1$ for $\lambda = (\lambda_1, \ldots, \lambda_n) \in \Lambda$ defines an algebraic subgroup of $\mathbb{G}_m^n$. Every algebraic subgroup of $\mathbb{G}_m^n$ is of this type. It is irreducible (as a variety) if and only if the lattice Λ is *primitive*; that is, $\Lambda = L_\Lambda \cap \mathbb{Z}^n$, where $L_\Lambda = \Lambda \otimes_{\mathbb{Z}} \mathbb{R} \subset \mathbb{R}^n$ is the linear span of Λ in $\mathbb{R}^n$.

An irreducible (as an algebraic variety) algebraic subgroup is called a *linear subtorus* (or just *subtorus*). If reducible, the identity component is a subtorus and every other component is a coset of it by a torsion point.

3.2 Definition ([85, p. 82]) A *torsion coset*, also known as a *special subvariety* of $\mathbb{G}_m^n$, is a translate of a subtorus by a torsion point.

Otherwise put, the torsion cosets are the irreducible components of algebraic subgroups; equivalently, they are the irreducible components of subvarieties defined by imposing some number of equations of the form $x^{a_1} \cdots x_n^{a_n} = 1$ with $(a_1, \ldots, a_n) \in \mathbb{Z}^n \setminus \{0\}$.

Observe that a torsion coset of positive dimension always contains an infinite number (and indeed a Zariski-dense set) of torsion points: if it has dimension k and one chooses k independent coordinates, one may set them to be arbitrary roots of unity and then solve for the other coordinates. As the equations are multiplicative, the remaining coordinates are also roots of unity. Also observe that the collection of torsion cosets is closed under taking irreducible components of intersections. Both these features hold in more general settings to be considered later. Specifically, they hold for the collection of special subvarieties of any (mixed) Shimura variety.

We can now state the generalization of Theorem 3.1, which can be formulated in several ways. One says that all the torsion points in V are accounted for by finitely many torsion cosets contained in V:

Given $V \subset \mathbb{G}_m^n$, there exist a finite number of torsion cosets $T_i \subset V$, $i = 1, \ldots, k$ such that every torsion point $x \in V$ is contained in T_i for some i.

Another is: the union $\cup T$ of all torsion cosets $T \subset V$ is a finite union. Or equivalently, the Zariski closure of any set of torsion points is a finite union of torsion cosets. We prove the following equivalent version.

3.3 Theorem ([329]; Multiplicative Manin–Mumford; MMM) *Let $V \subset \mathbb{G}_m^n$ be an algebraic subvariety. Then V contains only finitely many maximal torsion cosets.*

Theorem 3.3 is known in sharp effective forms (see some references in Remarks 3.14). The proof given here via point-counting adds no new information but is given in order to illustrate the strategy in its simplest incarnation.

The Original Manin–Mumford Conjecture

The classical Manin–Mumford conjecture, which inspired Lang's formulation in [324], is the statement in Theorem 3.3 with $\mathbb{G}_m^n$ replaced by an abelian variety (see after Theorem 6.7), where *torsion cosets* are cosets of abelian subvarieties by torsion points.

First proved in [451, 452], it has subsequently been reproved by a great variety of methods, see, for example, the survey [507], including a proof ([283]) using the model theory of difference fields. More recent approaches include the proof by point-counting in [440] along the lines of the proof given next, and a proof ([449]) using perfectoid spaces.

Proof of MMM via Point-Counting

The proof uses the uniformization of $\mathbb{G}_m^n$ by the (modified) exponential function

$$e : \mathbb{C}^n \to \left(\mathbb{C}^\times\right)^n, \quad e(z) = e(z_1, \ldots, z_n) = (e^{2\pi i z_1}, \ldots, e^{2\pi i z_n}).$$

The torsion points of $\left(\mathbb{C}^\times\right)^n$ are precisely the images of rational points in $\mathbb{C}^n$. The theorem may be reformulated as asserting that the rational points in $e^{-1}(V)$ are contained in the $\mathbb{Z}^n$ translates of a *finite* number of rational linear subvarieties that are contained in $e^{-1}(V)$.

We identify $\mathbb{C}$ with $\mathbb{R}^2$ using real and imaginary parts. The proof involves an application of the Counting Theorem to a suitable definable subset of $\mathbb{R}^{2n}$, but to turn the conclusion of the Counting Theorem (that there are few rational points outside the algebraic part) into the above finiteness statement requires various additional ingredients. These are partly of an arithmetic nature and partly of a geometric nature, to characterize the algebraic part in suitable terms.

Let us first observe that we may assume the subvariety V is defined over a number-field. Indeed, on any algebraic variety V defined over $\mathbb{C}$, the Zariski closure of its algebraic points is a finite union of subvarieties defined over finite extensions of $\mathbb{Q}$. Since torsion points are algebraic, they belong to this union, and so the theorem holds for V if it holds for each of these subvarieties.

The set

$$F = [0, 1) \times \mathbb{R} \subset \mathbb{C}$$

is a fundamental domain for the action of $\mathbb{Z}$ on $\mathbb{C}$ by translation, under which the map $z \mapsto e(z)$ is invariant. We set

$$Z = e^{-1}(V) \cap F^n$$

and gather some ingredients (which map roughly to the ingredients required to prove the André–Oort conjecture as set out in [509]).   □

Definability

3.4 Definability Ingredient   *The set $Z \subset \mathbb{R}^{2n}$ is definable in an o-minimal structure.*

See Definition 8.3 and Theorem 8.26 and the sequel. Specifically, Z is definable in the structure $\mathbb{R}_{\text{an, exp}}$. Note however that Z is *not* a set of the form of those in Proposition 2.7, all of which are definable in $\mathbb{R}_{\text{an}}$. The full graph of the real exponential function is not definable there. It is precisely to deal with uniformizations that have such exponential behaviour in the cusps, which are ubiquitous in the arithmetic contexts, that the greater generality of o-minimal structures is essential.

The full pre-image $e^{-1}(V)$ is not a definable set in any o-minimal structure, being a periodic repetition of Z. However, every point in $(\mathbb{C}^{\times})^n$ has a pre-image in the fundamental set F^n for the $\mathbb{Z}^n$ action, so it it is enough to work with Z.

Heights and Degrees

The complexity of a torsion point is conveniently measured by its order. As the Counting Theorem applies to definable sets in real Euclidean space, the height to be considered is of the relevant points in $\mathbb{R}^{2n}$.

Suppose $\zeta \in \mathbb{G}_{\text{m}}$ is a torsion point of (exact) order N, and let $z \in F \cap e^{-1}(\zeta)$. Then, in the real coordinates, $z = (0, q)$ or $(1, q)$ with $q = a/N \in \mathbb{Q}, (a, N) = 1$, and so $H(z) = N$.

3.5 Height Ingredient *The pre-image $z \in F^n$ of a torsion point ζ of order N is a rational point of height $H(z) \leq N$.*

While z is rational, ζ has quite high degree, bounded below by a positive power of its complexity. A root of unity ζ of exact order N has $[\mathbb{Q}(\zeta) : \mathbb{Q}] = \phi(N)$, where ϕ is the Euler totient function defined by

$$\phi(n) = \#\{a \in \mathbb{Z} : 1 \leq a \leq n - 1, \gcd(a, n) - 1\}.$$

(It is elementary that the roots of unity $e(a/N)$ for a with $\gcd(a, N) = 1$ are all conjugate.) And $\phi(N)$ is not much smaller than N: indeed, for every positive η,

$$\phi(N)/N^{1-\eta} \to \infty$$

as $N \to \infty$. This is a standard fact of elementary number theory; see, for example, [266, Theorem 327] (or [266, Theorem 328] for an even sharper result). Therefore we have the following.

3.6 Galois Ingredient *For every positive η there is a constant $c(\eta)$ such that*

$$[\mathbb{Q}(\zeta) : \mathbb{Q}] \geq c(\eta)N^{1-\eta}$$

for every root of unity ζ of (exact) order N.

Ingredients 3.4, 3.5, and 3.6 are deployed in conjunction with the Counting Theorem in the arithmetic part of the proof, in the following way. Assume, as we have done, that V is defined over a number-field. Suppose that V contains a torsion point ζ of some (sufficiently large, exact) order N. Taking $\eta = 1/2$ in 3.6, we see, in conjunction with 3.5, that

$$Z \text{ contains } \gg N^{1/2} \text{ rational points of height at most } N,$$

where the implied constant depends on the constants in 3.6 and the degree of the field of definition of V.

In view of 3.4, we may also apply the Counting Theorem with $\epsilon = 1/4$. It asserts that

$$Z - Z^{\text{alg}} \text{ contains } \ll N^{1/4} \text{ rational points of height at most } N.$$

(Note that the counting applies to rational points in $\mathbb{R}^2$, not just to rational points in $\mathbb{C}$, as the torsion points give, but this is of no significance.)

The upper and lower estimates are incompatible for sufficiently large N unless Z^{alg} is non-empty. We deduce (assuming the existence of such ζ) that Z contains positive-dimensional real semi-algebraic sets (which indeed must account for most of the rational points).

Since $e^{-1}(V)$ is a complex analytic set, we deduce that the real semi-algebraic subsets are contained in positive-dimensional complex algebraic sets $W \subset e^{-1}(V)$. This can be seen as follows. Take a non-singular point where A may be parameterized by smooth real algebraic functions $\phi = (\phi_1, \ldots, \phi_n)$ of some real parameters $t = (t_1, \ldots, t_m)$. Now consider the parameterizing functions for complex values of the parameters. If $F_1, \ldots, F_m$ are polynomials defining V, then $F\big(e(\phi(t))\big) = 0$ for all real parameter values, and hence vanishes for all complex values, as the zeros of complex functions are isolated. Thus, $e^{-1}(V)$ contains a complex algebraic set W containing A, or, otherwise put, we have a complex algebraic $W \subset \mathbb{C}^n$ with $e(W) \subset V$.

By this step, the problem about rational points has become a geometric problem about the uniformizing function $e(z)$, and more precisely a *functional transcendence* problem: when can we have $e(W) \subset V$?

Ax–Lindemann

We have already mentioned (on page 4) that this can only happen in particular circumstances governed by the functional analogue of Schanuel's conjecture, a 1971 result of Ax [20] known as Ax–Schanuel. This theorem with its analogues is discussed in detail in Part III.

The obvious way it can happen is if $W \subset L$ for some translate L of a proper rational linear subspace of $\mathbb{C}^n$. Then, $e(L) = T$ is a coset of a proper subtorus, and one could have $T \subset V$. Ax–Schanuel implies, as a special case, that this is the only way it can happen.

3.7 Definition We refer to translates of rational linear subspaces of $\mathbb{C}^n$ as *weakly special subvarieties* of $\mathbb{C}^n$; their images are cosets of subtori of $\mathbb{G}_m^n$, which we also call *weakly special subvarieties*.

3.8 Theorem ([20]) *Let $L \subset \mathbb{C}^n$ be a weakly special subvariety with image $T = e(L) \subset \mathbb{G}_m^n$. Let $W \subset L$ be an algebraic subvariety. Then, $e(W)$ is Zariski-dense in T unless W is contained in a proper weakly special subvariety of L.*

One special case of Schanuel's conjecture is the classical Lindemann, or Lindemann–Weierstrass Theorem.

3.9 Theorem (Lindemann(–Weierstrass) Theorem; see, for example, [24, 325, 388]) *Let $\alpha_1, \ldots, \alpha_n$ be algebraic numbers. Then, $e^{\alpha_1}, \ldots, e^{\alpha_n}$ are algebraically independent over $\mathbb{Q}$ unless $\alpha_1, \ldots, \alpha_n$ are linearly dependent over $\mathbb{Q}$.*

So if $\alpha = (\alpha_1, \ldots, \alpha_n) \in \overline{\mathbb{Q}}^n$, then $e^\alpha = (e^{\alpha_1}, \ldots, e^{\alpha_n})$ is $\overline{\mathbb{Q}}$-Zariski-dense in $\mathbb{C}^n$ unless α lies in a proper $\mathbb{Q}$ subspace. Theorem 3.8 is thus a functional analogue of Lindemann's Theorem, hence the retronym "Ax–Lindemann" for this special case of Ax–Schanuel.

Returning to $W \subset \mathbb{C}^n$ as above, we may let L be the smallest translate of a rational linear space containing W (just take the intersection of all translates of rational subspaces containing W), and put $T = e(L)$. If $e(W) \subset V$, then we must have $T \subset V$, since $e(W)$ is Zariski-dense in T. We may therefore formulate Theorem 3.8 equivalently as follows.

3.10 Ax–Lindemann Ingredient *A maximal algebraic subvariety $W \subset e^{-1}(V)$ is weakly special.*

A Finiteness Property

The last ingredient we need is a certain finiteness property: the maximal cosets contained in V are translates of *finitely many* subtori.

3.11 Finiteness Ingredient *Given $V \subset \mathbb{G}_m^n$, there are finitely many subtori $H^{(1)}, \ldots, H^{(k)} \subset \mathbb{G}_m^n$ such that every coset of a subtorus that is contained in V, and maximal among such cosets contained in V, is a coset of one of the $H^{(i)}$.*

A direct effective proof that the number of $H^{(i)}$ is bounded by a constant depending only on n and the degree of the polynomials defining V is given in [94]. An explicit statement is in [85, Theorem 3.3.9(b)]. Alternatively, in view of

3.10, it also follows immediately (though ineffectively) by o-minimality (such a proof is given in §8.36), or by model-theoretic compactness. The corresponding result for abelian varieties was proved earlier in [84].

Now, given one of these maximal tori $H = H^{(i)}$, we may ask: which translates xH of H have $xH \subset V$? We follow the treatment in [85, Theorem 3.3.9], though in a simplified way. Let $\Lambda \subset \mathbb{Z}^n$ be the lattice of exponent vectors corresponding to H. If $\dim H = r$, then Λ has rank $n - r$ and is primitve, as H is a subtorus. We can take $n - r$ exponent vectors $\lambda_1, \ldots, \lambda_{n-r} \in \mathbb{Z}^n$ forming a basis for Λ, and complete it to a basis $\lambda_1, \ldots, \lambda_n$ of $\mathbb{Z}^n$. The monoidal transformation $\phi : \mathbb{G}_m^n \to \mathbb{G}_m^n$ given by

$$x = (x_1, \ldots, x_n) \mapsto x^\lambda = (x^{\lambda_1}, \ldots, x^{\lambda_n})$$

is invertible. By means of this map, cosets of H are parameterized by points in $\mathbb{G}_m^{n-r}$. Namely, $H = \phi^{-1}(1_{n-r} \times \mathbb{G}_m^r)$, and cosets of H are of the form $\phi^{-1}(\{y\} \times \mathbb{G}_m^r)$ for $y \in \mathbb{G}_m^{n-r}$. It is an easy check that if $z \in H$, then $\phi(zy) = \phi(y)$; note that the notation hides the dependence on ϕ.

3.12 Definition   By a *family of weakly special subvarieties* in $\mathbb{G}_m^n$ we mean the family of cosets of some subtorus H parameterized as above (with some choice of vectors to complete the basis Λ giving the map ϕ) by some $Y = \mathbb{G}_m^{n-r}$. We suppress mention of the particular choices. Formally, the family $T = T(H, \Lambda)$ is thus the graph of $\phi : \mathbb{G}_m^n \to Y$, viewed as the collection of fibres $T_y \subset \mathbb{G}_m^n$. The total space of such a family is always $\mathbb{G}_m^n$.

For each $H^{(i)}$ we define $V^{(i)} \subset \mathbb{G}_m^{n-r}$ by

$$V^{(i)} = \{y \in \mathbb{G}_m^{n-r} : H_y^{(i)} \subset V\}.$$

Then, $V^{(i)} \subset \mathbb{G}_m^{n-r}$ is a closed subvariety [85, Theorem 3.3.9(c)]. Cosets of subtori contained in $V^{(i)}$ correspond to cosets of subtori containing $H^{(i)}$ that are contained in V. In particular, torsion cosets of $H^{(i)}$ that are maximal special subvarieties of V correspond to torsion points of $V^{(i)}$ that are not contained in any larger special subvariety contained in $V^{(i)}$.

Finally, we can assemble all the above ingredients and observations to prove MMM.

3.13 Proof of MMM   We prove the theorem by induction on the ambient dimension n, the conclusion being trivial when $n = 1$.

Let $V \subset \mathbb{G}_m^n$ and assume the theorem holds for subvarieties of $\mathbb{G}_m^m$ whenever $m < n$. We may assume that V is defined over a number-field. Let

$$H^{(1)}, \ldots, H^{(k)}$$

be the finite list of subtori afforded by Ingredient 3.11. If $H^{(i)}$ has positive dimension then, by induction, $V^{(i)}$ contains only finitely many special points as maximal special subvarieties. Therefore, V contains only finitely many maximal special subvarieties that are translates of $H^{(i)}$, and hence only finitely many of positive dimension. Let V^* be the union of these. Let K be a number-field that is a field of definition for V and V^*.

Suppose $P \in V$ is a torsion point not contained in V^*, of (exact) order N. The conjugates of P over K are again torsion points in $V \backslash V^*$. If N is sufficiently large, then, comparing the lower bound of 3.6 and the height upper bound of 3.5 with the upper bound of the Counting Theorem, we see that the definable (by 3.4) set Z contains a positive-dimensional semi-algebraic set and hence a positive-dimensional complex algebraic variety, which moreover contains the pre-images of special points.

By Ax–Lindemann (3.10), $e^{-1}(V)$ contains a weakly special subvariety containing the above algebraic subvariety, hence of positive dimension, whose image in V contains some of the conjugates of P, and so is special. This contradicts the fact that all the positive-dimensional special subvarieties of V are accounted for by V^*.

Hence N is bounded.   □

Observe how the combination of the arithmetic properties of the torsion points and the geometric (functional transcendence) properties of the uniformization enables the conclusion of the Counting Theorem to be leveraged into a finiteness statement.

3.14 Remarks

1. Theorem 3.3 is known in precise forms, see, for example, [348]. See also [283] for explicit results in the semi-abelian case, in which finiteness was proved in [278]. The results of [342] and [157] are explicit; these are generalized and strengthened in [195].

2. Theorem 3.3 is strengthened by the Bogomolov conjecture on small points proved in [546] (see also [94]), and the equidistribution of small points ([62]); both have famous analogues for abelian varieties, on which see also [321].

4

Powers of the Modular Curve as Shimura Varieties

Towards the Statement of Modular André–Oort

The André–Oort conjecture is an analogue of the Manin–Mumford conjecture for Shimura varieties ([374, 375]); see also Chapter 6. The simplest Shimura variety is the modular curve $Y(1)$, the moduli space of complex elliptic curves, which may be identified with the affine line $\mathbb{A}^1(\mathbb{C}) = \mathbb{C}$. The background is recalled below. However, the André–Oort conjecture is trivial for one-dimensional Shimura varieties, and to get an interesting statement one needs to look to higher-dimensional ones, the simplest being the cartesian powers of the modular curve.

A Shimura variety has a collection of special subvarieties. Among them are the special points, which are the special subvarieties of dimension zero, and which are Zariski-dense in every special subvariety. The André–Oort conjecture addresses the converse: which subvarieties of a Shimura variety contain a Zariski-dense set of special points?

The purpose of this chapter is to give a brief introduction to the modular curve $Y(1)$ and its arithmetic, and then to define in some detail the collection of special subvarieties (and the larger collection of weakly special subvarieties) in cartesian powers $Y(1)^n$. With these we can state the modular case of the André–Oort conjecture, and give a proof of it, in the next chapter.

The Modular Function and the Modular Curve

The special points and special subvarieties of $Y(1)^n$ arise from the arithmetic of the modular interpretation, which is reflected in properties of the modular function. In this sense, Modular André–Oort is the analogue of MMM (Theorem 3.3) obtained by replacing the cartesian powers of the (modified) exponential function

$$e \colon \mathbb{C}^n \to \mathbb{G}_{\mathrm{m}}^n$$

by the cartesian powers of the modular function (see page 3 and Definition 4.1 below)

$$j \colon \mathbb{H}^n \to Y(1)^n,$$

and replacing the torsion cosets of $\mathbb{G}_{\mathrm{m}}^n$ by the special subvarieties of $Y(1)^n$. Note that $\mathbb{G}_{\mathrm{m}}^n$ is not a Shimura variety, though it is a mixed Shimura variety; see around Definitions 18.4 and 20.21.

Other modular curves are finite covers of $Y(1)$, and the André–Oort conjecture for products of modular curves follows from the result for $Y(1)^n$. So we say little about them beyond giving the definition and a bit of detail about the modular curve $Y(2)$, which we encounter later in conjunction with the Legendre family of elliptic curves.

Our main reference for the following background material is the survey [541], though there are many other sources.

4.1 Definition   We have already introduced (on page 3) the *modular function* or *j-function* given by

$$j(z) = \frac{E_4(z)^3}{\Delta(z)} = \frac{1}{q} + 744 + \sum_{n=1}^{\infty} c_n q^n, \quad \text{where } q = e(z),$$

holomorphic in the complex upper half-plane, mapping onto the complex affine line.

The modular function parameterizes elliptic curves. Recall that an *elliptic curve* E defined over $\mathbb{C}$ may be uniformized by the complex numbers. The uniformization is doubly periodic, with some lattice of periods $\Lambda \subset \mathbb{C}$. Thus, $E = \Lambda \backslash \mathbb{C}$. We do not go into the underlying theory of doubly periodic functions and the construction of modular functions. Scaling the lattice does not change the quotient elliptic curve, up to isomorphism, as an algebraic variety over $\mathbb{C}$, so one may normalize the lattice to be of the form $\Lambda = \Lambda_z = \mathbb{Z} + \mathbb{Z}.z$ for some $z \in \mathbb{H}$.

Then, $j(z)$ is the *j-invariant* of the elliptic curve $E = \Lambda_z \backslash \mathbb{C}$. The *j*-invariant of an elliptic curve given, say, by a Weierstrass equation

$$E : y^2 = x^3 + Ax + B, \quad A, B \in \mathbb{C}, \quad \Delta(E) = 4A^3 + 27B^2 \neq 0$$

is given by

$$j(E) = \frac{1728(4A^3)}{4A^3 + 27B^2}$$

(here $\Delta(E)$ is the *discriminant* of E).

Two elliptic curves over $\mathbb{C}$ are isomorphic over $\mathbb{C}$ if and only if they have the same j-invariant, which may be any complex number.

4.2 Definition   The *modular curve* $Y(1) = Y(1)(\mathbb{C}) = \mathbb{C}$ parameterizes elliptic curves over $\mathbb{C}$, up to isomorphism over $\mathbb{C}$, by their j-invariant.

$SL_2(\mathbb{Z})$-Invariance

Changing the basis of the lattice and rescaling it does not change the elliptic curve, up to isomorphism over $\mathbb{C}$. This amounts to sending the lattice basis $\{1, z\}$ to some other basis $\{cz + d, az + b\}$, where $a, b, c, d \in \mathbb{Z}$ with $ad - bc = \pm 1$, and then rescaling to the form $\{1, z'\}$. The requirement that $z' \in \mathbb{H}$ entails $ad - bc = +1$, and we have the action of $SL_2(\mathbb{Z})$ on $\mathbb{H}$ by *fractional linear transformations*

$$z' = \gamma z = \frac{az + b}{cz + d}, \quad \gamma = \begin{pmatrix} a & b \\ c & d \end{pmatrix} \in SL_2(\mathbb{Z}).$$

As $j(z)$ is the j-invariant of $\Lambda_z \backslash \mathbb{C}$, it is invariant under the action of $SL_2(\mathbb{Z})$. The action has the following well-known *fundamental domain*:

$$F = \{z = x + iy \in \mathbb{H} : -1/2 \le \mathrm{Re}(z) \le 1/2, |z| \ge 1\}.$$

Actually this is not quite a true fundamental domain, as the modular group of transformations of $\mathbb{H}$ includes $z \mapsto z + 1$, which identifies the line $z = -1/2$ with the line $z = 1/2$, while $z \mapsto -1/z$ identifies the left and right halves of the circular part of the boundary. However this makes no difference for us, and the true fundamental domain formed by removing the superfluous half-boundary is, like F above, a semi-algebraic set in the real coordinates on $\mathbb{C}$ provided by real and imaginary parts.

The j-function takes each complex value exactly once in the (true) fundamental domain.

4.3 Proposition ([541, p. 22])   *The j-function is invariant under the action of $SL_2(\mathbb{Z})$ on $\mathbb{H}$ and identifies $SL_2(\mathbb{Z}) \backslash \mathbb{H}$ with $Y(1)$. In particular, for $z, w \in \mathbb{H}$,*

$$j(z) = j(w) \text{ if and only if } \exists \gamma \in SL_2(\mathbb{Z}) : w = \gamma z.$$

Modular Polynomials

Modular polynomials track the existence of *isogenies* (homomorphisms with finite kernels) between elliptic curves. For example, if there is an isogeny $\phi \colon E \to E'$ of degree 2, then, under some uniformization of E with lattice

$\Lambda = \mathbb{Z} + z\mathbb{Z}$, the pre-image of $0 \in E'$ is a lattice $\Lambda' = \mathbb{Z} + z'\mathbb{Z}$ corresponding to E' that contains Λ with index 2, so that $z' \in \{z/2, (1 + z)/2, 2z\}$.

The condition on the j-invariants that the corresponding curves admit an isogeny of given degree is algebraic. Accordingly, the functions $j(z)$ and $j(2z)$ are algebraically dependent. The relation is given by $\Phi_2\big(j(z), j(2z)\big) = 0$, where ([541, p. 70])

$$\Phi_2(X, Y) = X^3 - X^2Y^2 + Y^3$$
$$+ 1488X^2Y + 1488XY^2 - 162000X^2 + 40773375XY - 162000Y^2$$
$$+ 8748000000X + 8748000000Y - 157464000000000.$$

More generally, any $g \in \mathrm{GL}_2^+(\mathbb{R})$ (where "+" indicates positive determinant), acts on $\mathbb{H}$ by

$$g = \begin{pmatrix} a & b \\ c & d \end{pmatrix}, \quad gz = \frac{az + b}{cz + d}, \quad z \in \mathbb{H}.$$

If $g \in \mathrm{GL}_2^+(\mathbb{Q})$, one can scale it to have integer entries (which does not change the action) having $\gcd(a, b, c, d) = 1$, and then set $N = N(g) = \det(g)$. We then have the following result.

4.4 Proposition ([541, Proposition 23, p. 68, and Remark p. 70]; or [325, Ch. 5]) *For $g \in \mathrm{GL}_2^+(\mathbb{Q})$ and $N = N(g)$ as above, the functions $j(z)$ and $j(gz)$ are algebraically dependent. Specifically, we have*

$$\Phi_N\big(j(z), j(gz)\big) = 0,$$

where Φ_N are the modular polynomials. The degree of Φ_N is given by the Dedekind ψ function $\psi(N) = N \prod_{p|N} \left(1 + \frac{1}{p}\right)$.

The *modular polynomials* $\Phi_N(X, Y)$ lie in $\mathbb{Z}[X, Y]$. They are symmetric for $N \geq 2$; $\Phi_1(X, Y) = X - Y$. They detect the existence of an isogeny of degree N with cyclic kernel between the corresponding elliptic curves. Their symmetry registers the fact that if there is a degree N cyclic isogeny $E \to E'$, then there is also such an isogeny $E' \to E$ (the *dual isogeny*). Note that the multiplication map $n : E \to E$ has kernel isomorphic to $\mathbb{Z}/n\mathbb{Z} \times \mathbb{Z}/n\mathbb{Z}$ and hence is not cyclic for $n \neq \pm 1$.

Note also that as the image of $\{(z, gz) : z \in \mathbb{H}\}$ under the uniformization $j \colon \mathbb{H}^2 \to \mathbb{C}^2$, the modular curve

$$T_N = \{(x_1, x_2) : \Phi_N(x_1, x_2) = 0\}$$

is an irreducible (over $\mathbb{C}$) algebraic curve. More generally, if one takes $g_1, \ldots, g_k \in \mathrm{GL}_2^+(\mathbb{Q})$, then the image in $Y(1)^k$ of

$$z \mapsto \big(j(g_1z), \ldots, j(g_kz)\big), \quad z \in \mathbb{H}$$

is an irreducible algebraic curve. It is a component of the intersections of the corresponding modular hypersurfaces

$$\Phi_{N(g_i g_j^{-1})}(x_i, x_j) = 0, \quad 1 \le i < j \le k.$$

Observe further that the curve parameterized by $z \mapsto (g_1 z, \ldots, g_k z), z \in \mathbb{H}$, is unchanged if it is reparameterized by replacing $g_i z$ by $g_i h z$ for any $h \in SL_2(\mathbb{R})$. It is really the "ratios" $g_i g_j^{-1}$ as a rational point in a "projective" $SL_2(\mathbb{R})^n$ that matter; the image in $Y(1)^k$ is unchanged if one replaces g_i by $\gamma_i g_i, \gamma_i \in SL_2(\mathbb{Z})$, and it is then the *slope* (defined in Definition 4.14) that is preserved.

Singular Moduli

Endomorphisms of an elliptic curve $E = \Lambda_z \backslash \mathbb{C}$ correspond to complex numbers μ that preserve the lattice; that is, $\mu \Lambda_z \subset \Lambda_z$. Generically, only integers preserve the lattice, but it is an elementary fact that there is non-integer *complex multiplication* if and only if z is quadratic over $\mathbb{Q}$. The corresponding elliptic curve is said to have *complex multiplication* (CM).

4.5 Definition

1. A *singular modulus* or *CM point* is a complex number $\sigma = j(z)$, where $z \in \mathbb{H}$ is quadratic over $\mathbb{Q}$; that is, $[\mathbb{Q}(z) : \mathbb{Q}] = 2$.

2. A quadratic $z \in \mathbb{H}$ satisfies a *minimal polynomial* over $\mathbb{Z}$ of the form

$$az^2 + bz + c = 0, \quad a, b, c \in \mathbb{Z}, \quad \gcd(a, b, c) = 1$$

   and one defines the *discriminant* $D(\sigma) = b^2 - 4ac$ (which is negative as z is not real). The corresponding elliptic curve has CM by (i.e. its ring of endomorphisms is isomorphic to) the unique quadratic order O_D of discriminant D.

3. We define the *complexity* of a singular modulus σ of discriminant D to be

$$\Delta(\sigma) = |D(\sigma)|.$$

An elliptic curve with CM has non-trivial cyclic isogenies with itself. Consequently, if σ is a singular modulus, there exists N such that $\Phi_N(\sigma, \sigma) = 0$. In fact, there exist infinitely many such N. Therefore, singular moduli are algebraic numbers. A closer analysis shows they are in fact algebraic integers.

4.6 Proposition ([541, Corollary p. 71])   *Singular moduli are algebraic integers.*

Furthermore, according to the classical theory of complex multiplication initiated by Kronecker, singular moduli generate abelian extensions of the corresponding imaginary quadratic field. We will not say much about the rich subject of the arithmetic of singular moduli, but note the following key consequence.

4.7 Proposition ([541, Proposition 25]) *Let σ be a singular modulus of discriminant $D = D(\sigma)$. Then, $[\mathbb{Q}(\sigma) : \mathbb{Q}] = h(D) = h(O_D)$, the class number of the corresponding quadratic order.*

Special Subvarieties of $Y(1)$

The variety $Y(1) = Y(1)(\mathbb{C}) = \mathbb{A}^1(\mathbb{C}) = \mathbb{C}$ is the simplest Shimura variety (though technically singular moduli are Shimura varieties, of dimension 0).

4.8 Definition

1. The *special points* in $Y(1)$ are the singular moduli.
2. The *special subvarieties* of $Y(1)$ are the special points and $Y(1)$ itself.

The special points comprise a countably infinite set of algebraic points. They are the analogues of torsion points in $\mathbb{G}_\mathrm{m}$.

Special Subvarieties of $Y(1)^2$

The variety $Y(1)^2$ parameterizes pairs of elliptic curves. Generically, neither curve is CM, and there is no isogeny between them. For either of these things to happen is special.

4.9 Definition The *special subvarieties* of $Y(1)^2$ are (taking (x_1, x_2) as coordinates):

1. $Y(1)^2$ itself.
2. The lines $x_1 = \sigma$, where σ is a singular modulus.
3. The lines $x_2 = \sigma$, where σ is a singular modulus.
4. The modular curves $T_N = \{(x_1, x_2) : \Phi_N(x_1, x_2) = 0\}$, where $N \geq 1$.
5. The points (σ_1, σ_2), where σ_1, σ_2 are singular moduli (the *special points*).

If $z \in \mathbb{H}$ is quadratic, then so is gz for any $g \in \mathrm{GL}_2^+(\mathbb{Q})$; therefore, modular curves contain infinitely many (and indeed an analytically dense set of) special points.

The intersection of two modular curves $\Phi_N(x_1, x_2) = 0$ and $\Phi_M(x_1, x_2) = 0$ consists of special points. To see this, we may consider curves $z \mapsto (z, gz)$, $z \mapsto (z, hz)$ with $N(g) = N, N(h) = M$ which, under the modular function, parameterize the two curves. At an intersection point (x_1, x_2) we must have $x_1 = j(z), x_2 = j(gz) = j(h\gamma z)$ for some $\gamma \in \mathrm{SL}_2(\mathbb{Z})$, and then z is a fixed point of an element of $\mathrm{GL}_2^+(\mathbb{Q})$, and quadratic.

Thus, the irreducible components of the intersections of special subvarieties are again special.

The special subvarieties are the images in $Y(1)^2$ of suitable loci in $\mathbb{H}^2$: for $g \in \mathrm{GL}_2^+(\mathbb{Q})$, the image under j of $\{(z, gz) : z \in \mathbb{H}\}$ is the modular curve $T_N, N = N(g)$.

Special and Weakly Special Subvarieties of $Y(1)^n$

The variety $Y(1)^n$ parameterizes n-tuples of elliptic curves, up to isomorphism over $\mathbb{C}$. For a generic point in $Y(1)^n$, none of the corresponding elliptic curves is CM (i.e. special), and there are no isogenies between them. Either condition is special. One also wants the components of intersections of special subvarieties to be special, and these conditions determine the collection.

4.10 Definition (Equivalent to [198, Definition 1.1])   Let $n \geq 1$. The *special subvarieties* of $Y(1)^n$ are defined as follows:

1. The variety $Y(1)^n$ is special.
2. For $i \neq j$ and $N \geq 1$, the subvariety defined by $\Phi_N(x_i, x_j) = 0$ is special.
3. For $1 \leq i \leq n$ and a singular modulus σ, the subvariety $x_i = \sigma$ is special.
4. If $S, T \subset Y(1)^n$ are special, then the irreducible components of $S \cap T$ are special.
5. A special subvariety of dimension zero is called a *special point*.

As the intersection points of modular curves are singular moduli, the special points in $Y(1)^n$ are precisely the n-tuples of singular moduli.

A special subvariety of $Y(1)^n$ is thus an irreducible component of the variety defined by the application of *some* modular polynomials applied to *some* pairs of distinct coordinates, and by setting *some other* coordinates to be constant (though this can also be accomplished by imposing a modular condition $\Phi_N(x_i, x_i) = 0$).

There is a larger class of *weakly special subvarieties*. While not mentioned in the theorem, these are an essential part of the picture and of the proof.

4.11 Definition   The *weakly special subvarieties* of $Y(1)^n$ are defined as follows:

1. The variety $Y(1)^n$ is weakly special.
2. For $i \neq j$ and $N \geq 1$, the subvariety defined by $\Phi_N(x_i, x_j) = 0$ is weakly special.

3. For $1 \leq i \leq n$ and a complex number c, the subvariety $x_i = c$ is weakly special.

4. If $S, T \subset Y(1)^n$ are weakly special and $U \subset_{\mathrm{cpt}} S \cap T$, then U is weakly special.

4.12 Definition   The irreducible components of the pre-images in $\mathbb{H}^n$ of special (respectively weakly special) subvarieties are called *special subvarieties* (respectively *weakly special subvarieties*) of $\mathbb{H}^n$.

Note that it is not standard to refer to special subvarieties of $\mathbb{H}^n$, and they are sometimes referred to by some variant or other terminology, such as "pre-special" or "co-special". It is convenient for us to use the same term for both.

We need a more detailed description in the course of proving Modular André–Oort, and for further use in Part IV.

Detailed Description of Special and Weakly Special Subvarieties

Imposing some collection of modular relations and fixing some coordinates partitions the coordinates into the subset (possibly empty) of coordinates that are constant, and the sets of coordinates (possibly singletons) that are non-constant but related to each other by modular relations. On each set of related non-constant coordinates, the special subvariety is the image of a curve of the form

$$z \mapsto (g_1 z, \ldots, g_k z),$$

where $g = (g_1, \ldots, g_k) \in \mathrm{GL}_2^+(\mathbb{Q})^k$.

Thus, special subvarieties in $Y(1)^n$ are the images in $Y(1)^n$ of products of images of maps of the form $z \mapsto (g_1 z, \ldots, g_k z), z \in \mathbb{H}$, where $g_i \in \mathrm{GL}_2^+(\mathbb{Q})$, and special points. The images in $Y(1)^k$ of images of maps of the form $z \mapsto (g_1 z, \ldots, g_k z), z \in \mathbb{H}$, are components of the intersection of suitable modular curves on pairs of coordinates.

Special points are the images of quadratic points. Special subvarieties have dense sets of special points because $\mathrm{GL}_2^+(\mathbb{Q})$ preserves quadratic points.

It is convenient to refer to algebraic subvarieties of $\mathbb{H}^n$, although formally speaking there are none (except points): any positive-dimensional algebraic subvariety of $\mathbb{C}^n$ that intersects $\mathbb{H}^n$ in a non-empty set also escapes this set. We therefore enshrine the abuse of language in a definition.

4.13 Definition   By an *algebraic subvariety* of $\mathbb{H}^n$ we mean a complex-analytically irreducible component of the intersection $\mathbb{H}^n \cap W$, where $W \subset \mathbb{C}^n$ is an algebraic subvariety.

Thus, the images of the maps $z \mapsto (g_1 z, \ldots, g_k z), z \in \mathbb{H}$, and cartesian products of such images, are irreducible algebraic subvarieties of $\mathbb{H}^n$.

4.14 Definition

1. By a *strict partition* of $\{1, \ldots, n\}$ we mean a finite tuple $p = (p_0, p_1, \ldots, p_\ell)$ of disjoint subsets $p_i \subset \{1, \ldots, n\}$ whose union is $\{1, \ldots, n\}$ but in which p_0 (only) is permitted to be empty. The case $\ell = 0$ is permitted. For each i, we denote by $\mathbb{H}^{p_i}, Y(1)^{p_i}, \mathrm{GL}_2^+(\mathbb{Q})^{p_i}, \ldots$ the cartesian product over the indices in p_i.

2. Given a strict partition, certain suitable additional data determine a weakly special subvariety. The additional data consists of a $b_i \in \mathbb{H}$ for each $i \in p_0$, and a tuple

$$g^{(i)} \in \mathrm{GL}_2^+(\mathbb{Q})^{p_i}$$

for each $i \geq 1$. As only the ratios $g_i g_j^{-1}$ matter, it is convenient to fix a reference coordinate $j_i \in p_i$ with $i \geq 1$, for definiteness take the smallest indexed coordinate, and record the tuple as

$$g^{(i)} = (1, \ldots) \in \mathrm{GL}_2^+(\mathbb{Q})^{p_i}.$$

So if $\#p_i = 1$, we have $g^{(i)} = 1$.

3. These data determine a *weakly special subvariety in* $\mathbb{H}^n$, denoted $L(p, g)_b$, namely, the product of the point $b = (b_i, i \in p_0) \in \mathbb{H}^{p_0}$ and the images in $\mathbb{H}^{p_i}, i \geq 1$ of the maps $z \mapsto g^{(i)} z, z \in \mathbb{H}$. The image in $\mathrm{SL}_2(\mathbb{R})^{p - p_0}$ of the tuple $g = (g^{(1)}, \ldots, g^{(k)})$ is called the *slope* of the special subvariety $L(p, g)_b$.

4. If each $b_i, i \in p_0$ is quadratic, then $L(p, g)_b$ is a *special subvariety* of $\mathbb{H}^n$.

5. A *special subvariety* (respectively *weakly special subvariety*) in $Y(1)^n$ is the image under $j \colon \mathbb{H}^n \to Y(1)^n$ of a special subvariety (respectively weakly special subvariety) in $\mathbb{H}^n$. If $c = j(b)$, we denote this special subvariety $T(p, g)_c$. The data (and so the denotation) are not unique: the $\mathrm{GL}_2^+(\mathbb{Q})$-tuples can be changed by scaling without changing $L(p, g)_\tau$, but also the same fibre product of modular curves may be parameterized in different ways due to the invariance under $\mathrm{SL}_2(\mathbb{Z})$ (see the definition of slope next).

6. By the *slope* of a special subvariety $T(p, g)_\sigma$ we mean the slope of any pre-image of it, considered up to scaling, in

$$\prod_{i=1}^{k} \mathrm{SL}_2(\mathbb{Z})^{p_i - \{j_i\}} \backslash \mathrm{GL}_2^+(\mathbb{Q})^{p_i} / \mathrm{SL}_2(\mathbb{Z}),$$

where $\mathrm{SL}_2(\mathbb{Z})^{\{\ell\}}$ acts on $\mathrm{GL}_2(\mathbb{R})^{\{\ell\}}$ on the left and $\mathrm{SL}_2(\mathbb{Z})$ acts on all coordinates (except j_i) on the right, by g^{-1}. This is then well-defined.

7. We define the *complexity* of a special subvariety $T = T(p,g)_\sigma$ to be

$$\Delta(T) = \max\{\Delta(\sigma_\ell), N\}$$

over $\ell \in p_0$ and N such that $\Phi_N(x_i, x_j) = 0$ for $i, j \in p_k, k \geq 1, i \neq j$. This also is well-defined.

Families of Weakly Special Subvarieties

Weakly special subvarieties come in algebraic families where one allows the constant coordinates to vary. We formalize a definition as we need to work with these families.

4.15 Definition

1. Given just the data p, g, one has the *family* of weakly special subvarieties in $\mathbb{H}^n$, which we denote $L(p,g)$, and which comprises the weakly special subvarieties corresponding to an arbitrary choice of a tuple $b \in \mathbb{H}^{p_0}$.
2. Formally, we identify the family with its total space, which is the special subvariety consisting of the same group data on $p_1, \ldots, p_\ell$, but now letting the coordinates in p_0 to be non-constant and unrelated to each other (so they are added to the partition as singeltons and there are no constant coordinates). The weakly special subvarieties in the family are then the fibres of the projection map

$$L(p,g) \to \mathbb{H}^{p_0}.$$

The fibre over $z \in \mathbb{H}^{p_0}$ is denoted as $L(p,g)_z$.
3. There is a corresponding *family of weakly special subvarieties in* $Y(1)^n$, namely, the fibres of the projection

$$\phi: T(p,g) \to Y(1)^{p_0},$$

where $T(p,g)$, the image of $L(p,g)$ under j, is the total space of the family of weakly special subvarieties, and is special. The fibre over $x \in Y(1)^{p_0}$ is denoted $T(p,g)_x$.
4. Sometimes we may refer to the weakly special subvarieties in the family $L(p,g)$ as *translates* of $L(p,g)$, in analogy with the translates of rational linear subspaces in $\mathbb{C}^n$; likewise to fibres of $T(p,g)$ as translates of $T(p,g)$ in analogy with cosets of subtori of $\mathbb{G}_m^n$.

Some of the weakly special varieties in the family are special, namely, precisely those fibres $T(p,g)_x$ for which the point $x \in Y(1)^{p_0}$ is special. Likewise for families of weakly special subvarieties in $\mathbb{H}^n$. The fibres in a family all have the same dimension equal to the number of $p_i, i \geq 1$, and there is a unique family of weakly special subvarieties of dimension zero (i.e. points) given by $p = (p_0) = (\{1, \ldots, n\})$.

Strongly Special Subvarieties of $Y(1)^n$

Within the collection of special subvarieties are the strongly special ones, having no constant coordinates.

4.16 Definition   A *strongly special subvariety* in $\mathbb{H}^n$, respectively $Y(1)^n$, is a special subvariety in which no coordinate is constant.

Here the corresponding partition has $p_0 = \emptyset$, so each strongly special subvariety is the unique fibre in its family of weakly special subvarieties, and is special.

Möbius Subvarieties

We define a still larger collection of subvarieties in $\mathbb{H}^n$.

4.17 Definition   A *Möbius subvariety* of $\mathbb{H}^n$ is defined in the same way as a weakly special subvariety, starting with a strict partition $p = (p_0, p_1, \ldots, p_k)$ of $\{1, \ldots, n\}$, and a choice of a point $z_0 \in \mathbb{H}^{p_0}$, but with coordinates in the other $p_i, i \geq 1$ related by elements of $\mathrm{SL}_2(\mathbb{R})$ rather than $\mathrm{GL}_2^+(\mathbb{Q})$.

This is a semi-algebraic family (parameterized by copies of $\mathrm{GL}_2(\mathbb{R})^+$ and $\mathbb{H}^{p_0}$) of algebraic subvarieties of $\mathbb{H}^n$. Their images in $Y(1)^n$ are in general rather bad (generally Zariski-dense in $Y(1)^n$) unless the $\mathrm{SL}_2(\mathbb{R})$-tuples happen to be (up to scaling) elements of $\mathrm{GL}_2^+(\mathbb{Q})$.

Thus, the weakly special subvarieties correspond, modulo the scaling, to rational points in the $\mathrm{SL}_2(\mathbb{R})$ coordinates of these families.

Other Modular Curves

4.18 Definition (See e.g. [178, 1.2.1, 1.5])   A *modular curve* is the quotient of $\mathbb{H}$ by a *congruence subgroup* of $\mathrm{SL}_2(\mathbb{Z})$, where a congruence subgroup of $\mathrm{SL}_2(\mathbb{Z})$ is a subgroup containing the *principal congruence subgroup*

$$\Gamma(N) = \left\{ \begin{pmatrix} a & b \\ c & d \end{pmatrix} \in \mathrm{SL}_2(\mathbb{Z}) : \begin{pmatrix} a & b \\ c & d \end{pmatrix} \equiv \begin{pmatrix} 1 & 0 \\ 0 & 1 \end{pmatrix} \bmod N \right\}$$

for some $N \geq 1$.

In particular,

$$Y(2) = \mathbb{P}^1 \backslash \{0, 1, \infty\} = \Gamma(2) \backslash \mathbb{H}.$$

This modular curve is the moduli space of elliptic curves with *full level* 2 *structure* parameterizing elliptic curves together with a choice of basis of their 2-torsion points. One has the uniformization

$$\lambda: \mathbb{H} \to Y(2)$$

(see e.g. [2, Ch. 7, 3.4 and 3.5]). It is sometimes nicer to work with $Y(2)$ than $Y(1)$ because the covering $\mathbb{H} \to Y(2)$ is unramified. Thus, $\mathbb{H}$ is the universal cover of $Y(2)$, and $\Gamma(2)$, its fundamental group, is a free group on two generators.

The map $Y(2) \to Y(1)$ obtained by forgetting the level structure is given by

$$j = R(\lambda) = 2^8 \frac{(\lambda^2 - \lambda + 1)^3}{\lambda^2 (\lambda - 1)^2}$$

(see e.g. [287, Ch. 4, 1.4]), and the *special points* of $Y(2)$, which correspond to CM elliptic curves with level structure, are just the images of quadratic points under λ, and so are the pre-images of singular moduli under $j = R(\lambda)$.

5

Modular André–Oort

Statement

The simplest non-trivial case of the André–Oort conjecture is that of $Y(1)^2$ (or more generally a product of two modular curves), and is due to André [6]. This is the analogue of MMM for $\mathbb{G}_m^2$, Theorem 3.1 above, the theorem discussed in [324].

5.1 Theorem ([6, Théorème])   *Let $V \subset Y(1)^2$ be a curve. Then V contains only finitely many special points unless V is a special subvariety.*

Just as for MMM, there are several equivalent ways to formulate the André–Oort conjecture for $Y(1)^n$. We state the following version.

5.2 Theorem (Modular André–Oort (MAO); [427, Theorem 1.1])   *Let $V \subset Y(1)^n$. Then V contains only finitely many maximal special subvarieties.*

This theorem was proved by Edixhoven [198] under the assumption of GRH for imaginary quadratic fields (see also his earlier proof [196] of Theorem 5.1 under GRH). The unconditional proof in [427] is via point-counting.

Ingredients

The proof uses the uniformization of $Y(1)^n$ by the (cartesian power of the) modular function

$$j: \mathbb{H}^n \to \mathbb{C}^n.$$

The special points of $Y(1)^n$ are precisely the images of quadratic points in $\mathbb{H}^n$ (points each of whose coordinates is quadratic over $\mathbb{Q}$). The argument involves an application of the Counting Theorem for quadratic points to the definable set

$$Z = j^{-1}(V) \cap F^n,$$

where F is the classical fundamental domain (see Proposition 4.3 above) for the action of $SL_2(\mathbb{Z})$ on $\mathbb{H}$.

5.3 Definability Ingredient   *The set $Z \subset \mathbb{R}^{2n}$ is a definable set in an o-minimal structure; more specifically, in the structure $\mathbb{R}_{an,exp}$, see § 6.10 and Theorem 8.26.*

5.4 Height Ingredient   *The pre-image $z \in F$ of a special point $\sigma \in Y(1)$ is a quadratic point of height $H(z) \leq 2\Delta(\sigma)$.*

Proof   Let $aZ^2 + bZ + c$ be the minimal polynomial of z. So $\gcd(a, b, c) = 1$ and $D = b^2 - 4ac$. The condition that $z \in F$ is equivalent to the triple (a, b, c) being *reduced*, namely, $|b| \leq a \leq c$ and $b \geq 0$ if $a = |b|$ or $a = c$. Then, $4ac = b^2 - D(z) \leq ac - D(z)$, giving $3ac \leq |D(z)|$. In real coordinates we have $z = (u, v)$ with $u = -b/2a, v = \sqrt{|D|}/2a$ so that v is a root of the polynomial $4a^2V^2 - |D|$. Then, by [85, 1.6.5, 1.6.6], we get

$$H(u) \leq \max\{b, 2a\} \leq 2a \leq |D| = \Delta, \quad H(v) \leq \max\{4a^2, |D|\} \leq 2\Delta,$$

as required.   □

On the other hand, the number of conjugates of σ over $\mathbb{Q}$ is equal to the class number $h(D)$ of the corresponding quadratic order (see Proposition 4.7), which, by a famous (and famously ineffective) result of Siegel [491], is bounded below by $c(\epsilon)|D|^{1/2-\epsilon}$. Hence:

5.5 Galois Ingredient   *For every $\epsilon > 0$, we have $[\mathbb{Q}(\sigma) : \mathbb{Q}] \geq c_\epsilon \Delta(\sigma)^{1/2-\epsilon}$.*

The very slightly earlier, weaker result of Landau [323] (with exponent $1/8-\epsilon$), or indeed any positive power of $|D|$, would suffice. But known effective bounds, see, for example, [246], are too weak for the present purpose.

As special points are algebraic, we may (as with MMM) reduce to the case in which V is defined over a number-field, so that a fixed positive proportion of the Galois conjugates of a special point land back on V.

Thus, if V has a special point of sufficiently large complexity, then Z contains many quadratic points up to some height. It must therefore (by the Counting Theorem) contain positive-dimensional semi-algebraic sets and (by analytic continuation) complex algebraic subvarieties of $\mathbb{H}^n$ (in the sense of Definition 4.13).

To continue the argument we need to characterize the maximal algebraic subvarieties $W \subset j^{-1}(V)$. This is accomplished with the modular analogue of the Ax–Lindemann Theorem.

Ax–Lindemann for the Modular Function

As we have seen, the weakly special subvarieties $T(p, g)_c$ are the j-images of algebraic subvarieties $L(p, g)_b \subset \mathbb{H}^n$. The Modular Ax–Lindemann Theorem asserts that all occurrences of $j(W) \subset V$, where $W \subset \mathbb{H}^n$ is an algebraic subvariety in the sense of Definition 4.13 and $V \subset Y(1)^n$, are accounted for by weakly special subvarieties.

5.6 Theorem (Modular Ax–Lindemann; [427, Theorem 9.2]) *Let $L \subset \mathbb{H}^n$ be a weakly special subvariety and set $T = j(L)$. Let $W \subset L$ be an algebraic subvariety. Then, $j(W)$ is Zariski-dense in T unless W is contained in a proper weakly special subvariety of L.*

As before, the above form exhibits the analogy with transcendence properties, in this case conjectural, of the modular function. The following conjecture is implied by André's Generalized Grothendieck Period Conjecture; see [47, 48] and §13.6.

5.7 Conjecture (Modular Lindemann(–Weierstrass)) *Suppose $\alpha_1, \ldots, \alpha_n \in \mathbb{H}$ are algebraic numbers that are not quadratic. Then, $j(\alpha_1), \ldots, j(\alpha_n)$ are algebraically independent over $\mathbb{Q}$ unless there exist $k \neq \ell$ and $g \in \mathrm{GL}_2^+(\mathbb{Q})$ such that $\alpha_\ell = g\alpha_k$.*

For $n = 1$, this follows from a famous theorem of Schneider [478]; see, for example, [325, 388]. It remains open for $n \geq 2$.

5.8 Theorem ([478]) *One has both α and $j(\alpha) \in \overline{\mathbb{Q}}$ if and only if α is quadratic.*

Theorem 5.6 is equivalent to the following statement, just as Theorem 3.8 is equivalent to Ingredient 3.10; see Chapter 14.

5.9 Ax–Lindemann Ingredient *A maximal algebraic subvariety $W \subset j^{-1}(V)$ is weakly special.*

The assertion of Theorem 5.6 for an algebraic subvariety $W \subset \mathbb{H}^n$ can be expressed as follows. If $j(W)$ is not Zariski-dense in $\mathbb{C}^n$, then either some coordinate is constant on W or W is contained in $z_i = gz_j$ for some i, j, g. Note an interesting feature of this statement. If non-constant functions $j(z_i)$, with z_i restricted to an algebraic variety W, are algebraically dependent over $\mathbb{C}$, then it must be that two of them are dependent, by a modular relation.

The theorem also characterizes the *bi-algebraic subvarieties* for j: $\mathbb{H}^n \to$ $Y(1)^n$ (see Definition 6.3 below) as precisely the weakly special ones.

5.10 Corollary (Modular Ax–Hermite–Lindemann)   *If both $W \subset \mathbb{H}^n$ and $j(W)$ are algebraic, then W is weakly special (and conversely, if W is weakly special, then $j(W)$ is algebraic).*

In view of Schneider's Theorem one gets a further characterization of "bi-algebraic-over-$\overline{\mathbb{Q}}$" subvarieties.

5.11 Corollary   *If $W \subset \mathbb{H}^n$ is algebraic and defined over $\overline{\mathbb{Q}}$, and $j(W)$ is algebraic and defined over $\overline{\mathbb{Q}}$, then W is special (and conversely).*

Proof Suppose W and $j(W)$ are both algebraic and defined over $\overline{\mathbb{Q}}$. Then, W is weakly special. Further, any constant coordinates $z_i = c$ have the property that $c, j(c)$ are algebraic over $\mathbb{Q}$. Hence, they are special by Schneider's Theorem. The converse is immediate from the definitions. $\square$

Sketch Proof of Modular Ax–Lindemann

Much more about this theorem and its generalizations is in Part III. It follows from the Ax–Schanuel Theorem for Shimura varieties, Theorem 16.4, which is proved in §16.10. Here we briefly describe the main principle of the proof of this special case, which involves an independent application of point-counting.

5.12 Sketch Proof   Say $W \subset j^{-1}(V)$, with $W \cap F^n \neq \emptyset$. Every translate of W by $g \in SL_2(\mathbb{Z})^n$ is then also contained in $j^{-1}(V)$. On the one hand, we can show that many such translates again intersect F^n, this being equivalent to showing that W goes through many $SL_2(\mathbb{Z})^n$-translates of F. On the other hand, the whole space of $SL_2(\mathbb{R})^n$ translates of W, and hence also their intersections with $Z = j^{-1}(V) \cap F^n$, are definable families parameterized by $SL_2(\mathbb{R})^n$ (see Definition 8.6). We can definably select the parameters for which the intersection has given dimension (see Proposition 8.19). Hence, setting $k = \dim_{\mathbb{R}} W$ and

$$Y = \{g \in SL_2(\mathbb{R})^n : \dim_{\mathbb{R}} (gW \cap Z) = k\}$$

gives a definable set with many rational points. By the Counting Theorem and analytic continuation, this implies that the space of translates of W contained in $j^{-1}(V)$ contains positive-dimensional semi-algebraic sets. We then try to use these to make W bigger while remaining inside $j^{-1}(V)$ by taking the union over a positive-dimensional real algebraic family of translates of W and complexifying. But if W is maximal, as assumed in Ingredient 5.9, then W must be invariant under such translations, and we have $W = gW$ for many $g \in SL_2(\mathbb{Z})^n$.

With enough such invariance we show that W must be of the required form. This completes the sketch.

Observe that the appeal of this proof to the Counting Theorem is different to the one in opposing Galois bounds. Here it operates outward from the fundamental domain to the ambient space $\mathbb{H}^n$, with $\mathrm{SL}_2(\mathbb{Z})^n$-points as the rational points on some definable subset of $\mathrm{SL}_2(\mathbb{R})^n$, rather than inward on quadratic points of $Z \subset F^n$ arising from Galois conjugates.

A Finiteness Property

5.13 Finiteness Ingredient   *Given $V \subset Y(1)^n$, there is a finite set $T^{(i)} = T\big(p^{(i)}, g^{(i)}\big)$ of families of weakly special subvarieties such that every weakly special subvariety contained in V and maximal among weakly special subvarieties contained in V is a fibre of one of these families.*

This follows immediately by o-minimality once one has the Ax–Lindemann Theorem; a proof is given in Proposition 8.35, and indeed the same finiteness holds over any definable family of subvarieties V. It immediately implies finiteness for the number of strongly special subvarieties of V (even over V in a definable family). Ingredient 5.13 has been made effective (even explicit) in [70], using methods of differential algebra and Modular Ax–Lindemann. See more generally [73, 165].

Proof of MAO

Proof of Theorem 5.2   We prove the theorem by induction on the ambient dimension n, the conclusion being trivial for $n = 1$. Let $V \subset Y(1)^n$ and assume the conclusion holds for subvarieties of $Y(1)^m$ for all $m < n$. Since special points are algebraic, we may assume that V is defined over a number-field.

Let $T\big(p^{(1)}, g^{(1)}\big), \ldots, T\big(p^{(k)}, g^{(k)}\big)$ be the finite number of families of weakly special subvarieties afforded by 5.13. For each family $T\big(p^{(i)}, g^{(i)}\big)$, whose fibres have positive dimension, with parameter space $Y(1)^{p_0^{(i)}}$, set

$$V^{(i)} = \Big\{ x \in Y(1)^{p_0^{(i)}} : T\big(p^{(i)}, g^{(i)}\big)_x \subset V \Big\}.$$

Then, $V^{(i)} \subset Y(1)^{p_0^{(i)}}$ is a (closed) subvariety. By induction, it contains only finitely many maximal special subvarieties, among which are finitely many special points. These correspond to maximal special subvarieties of V that are fibres in the family $T\big(p^{(i)}, g^{(i)}\big)$.

Therefore, V contains only finitely many maximal special subvarieties of positive dimension. Let V^* be the union of these. Let K be a number-field over which V and V^* are defined.

Suppose $P \in V$ is a special point not contained in V^*, and let $\Delta(P) = \max \Delta(x_i)$ be its complexity. By the Galois Ingredient 5.5, choosing say $\epsilon = 1/4$, P has at least

$$\gg_d \Delta(P)^{1/4}$$

conjugates over K that lie again in $V \backslash V^*$. Each of these conjugates has a pre-image in Z that is a quadratic point of height $\ll \Delta(P)$ by 5.4; hence, by 5.3, if $\Delta(P)$ is sufficiently large, then Z must contain a positive-dimensional semi-algebraic set containing some (even many) of these conjugates, and by analytic continuation $j^{-1}(V)$ contains an algebraic subvariety of positive dimension containing some of these pre-images.

By the Ax–Lindemann Ingredient, 5.9, $j^{-1}(V)$ contains a weakly special subvariety of positive dimension, which is in fact special as it contains pre-images of special points. This is a contradiction as all the positive-dimensional special subvarieties of V are accounted for by V^*. Hence, $\Delta(P)$ is bounded. $\square$

6

Point-Counting and the André–Oort Conjecture

Shimura Varieties and the André–Oort Conjecture

The André–Oort conjecture is the same statement as MAO (Theorem 5.2) for a general *Shimura variety.*

Shimura varieties are central objects in arithmetic geometry and the theory of automorphic forms. They were introduced by Shimura (see e.g. [488, 489]) as generalizations of modular curves parameterizing families of abelian varieties with additional structure, and generalized by Deligne [171, 174] as varieties parameterizing certain Hodge structures. Introductions to the theory of Shimura varieties can be found in several places, in various levels of detail. See, for example, the introductory notes [374], or alternatively [378, 379], the brief introduction [375], an example-based introduction [322], and [163].

The prototype examples are the modular curves, and the paradigm examples are the *Siegel modular varieties* $\mathcal{A}_g$, which are described in further detail below.

6.1 Definition ([375], according to Shimura)   Let Ω be a Hermitian symmetric domain and denote by $\mathrm{Hol}(\Omega)^c$ the identity component of its group of holomorphic automorphisms. Take a semi-simple algebraic group G over $\mathbb{Q}$ with a surjection $G(\mathbb{R}) \to \mathrm{Hol}(\Omega)^c$ with compact kernel. A *(connected, pure) Shimura variety* X is a quotient

$$X = \Gamma \backslash \Omega,$$

where Γ is a torsion-free subgroup of $\mathrm{Hol}(\Omega)^c$ containing the image of a congruence subgroup of $G(\mathbb{Q})$ with finite index.

By Baily–Borel, the modular forms on Ω relative to Γ give X the structure of a quasi-projective variety, and provide a Γ-invariant uniformization

$$u \colon \Omega \to X.$$

44

The modern definition of a Shimura variety, following Deligne, starts with purely group-theoretic data (*Shimura datum*) comprising a reductive group G over $\mathbb{Q}$ and a $G(\mathbb{R})$-conjugacy class of homomorphisms $h\colon \mathbb{C}^\times \to G(\mathbb{R})$ having suitable properties. Taking a certain double-coset space gives a *Shimura variety* that is a finite union of connected Shimura varieties in the above sense. A Shimura variety is canonically defined over a canonical number-field, its *reflex field*. The Hermitian domain Ω is built out of the group data and can be realized, via the *Borel embedding*, as a semi-algebraic open subset of a projective algebraic variety $\widehat{\Omega}$, its *compact dual*. This gives an algebraic structure on Ω, and as the compact dual is defined over the reflex field, a $\overline{\mathbb{Q}}$-structure. We assume Ω is realized as a bounded symmetric domain in a suitable $\mathbb{C}^N$ via its *Harish-Chandra embedding*, except that, for $\mathcal{A}_g$, it is customary to take Ω in the unbounded realization given by Siegel upper half-space $\mathbb{H}_g$. However, any realization gives the same algebraic structure on Ω, see [309, Corollary B.2], and the $\overline{\mathbb{Q}}$-structures are the same in this case.

6.2 Definition By an *algebraic subvariety* $W \subset \Omega$ we mean a complex-analytically irreducible component of $\widehat{W} \cap \Omega$ for some algebraic subvariety $\widehat{W} \subset \widehat{\Omega}$.

Shimura varieties come in families where finite index $\Gamma' \subset \Gamma$ leads to finite covers $X' \to X$. By taking a suitable finite index subgroup one can always assume that X is smooth and the covering $u\colon \Omega \to X$ is unramified. The André–Oort conjecture (and more generally the Zilber–Pink conjecture) is insensitive to these finite coverings, so those conditions may be assumed when convenient, notwithstanding that, for example, the uniformization $\mathbb{H}_g \to \mathcal{A}_g$ of the *Siegel modular variety* (see below) is ramified.

6.3 Definition A Shimura variety X comes equipped with a collection of *special subvarieties* (also known as subvarieties of Hodge type); see, for example, [378, Definition 2.5], the zero-dimensional ones are called *special points*. There is a larger class of *weakly special subvarieties* (also known as *totally geodesic* subvarieties); see, for example, [378, §4.1].

Special subvarieties are essentially "sub-Shimura varieties" arising in a compatible way. There are countably infinitely many of them when X is positive-dimensional. A special subvariety contains a Zariski-dense (and even topologically dense) subset of special points. The weakly special subvarieties may be characterized ([512]; see Theorem 16.1) as the *bi-algebraic* subvarieties; that is, the algebraic subvarieties $W \subset X$ for which some (equivalently every) irreducible analytic component of $u^{-1}(W) \subset \Omega$ is algebraic. A subvariety is special if and only if it is weakly special and contains a special point ([378,

Theorem 4.3]). The collections of both special and weakly special subvarieties are closed under taking irreducible components of intersections.

6.4 Conjecture (The André–Oort conjecture (AO); [4, 394]) *Let X be a Shimura variety and V ⊂ X. Then V contains only finitely many maximal special subvarieties.*

One has equivalent formulations as for MMM: the usual one is that if $\Sigma \subset X$ is any set of special points, then an irreducible component of the Zariski closure of Σ is a special subvariety. Alternatively, a subvariety $V \subset X$ has a Zariski-dense set of special points if and only if it is a special subvariety.

AO was proposed as a problem by André [4, Ch. X §4.3, Problem 1] in the case of a curve in a general Shimura variety, and by Oort [394, §2] (see also [393, §6]) in the case of an arbitrary subvariety of $\mathcal{A}_g$. The context for André was a study of the exceptional fibres in a pencil of abelian varieties; that is, those fibres having a larger endomorphism ring than the generic fibre. He mentions the similarity with Coleman's conjecture [557, Conjecture 6] on Jacobians with CM. Oort mentions the possibility of formulating the analogue of his statement for Shimura varieties in general, which would be (equivalent to) the statement above. Both were mindful of the analogy with the Manin–Mumford conjecture (see [395] and [4, p. 216]).

Results

There are several surveys on the André–Oort conjecture; see [310, 391, 396, 535]. Here we briefly mention some of the main themes.

The basic case of $Y(1)^2$ is proved in [6] using methods of diophantine approximation. The conditional proof ([196]) under GRH uses Hecke operators and Galois-theoretic ideas, extended (under GRH) to Hilbert modular surfaces ([197]) and to arbitrary products of modular curves ([198]), and to products of two Shimura curves ([533]). Earlier, partial results in $\mathcal{A}_g$ (for sets of special points with suitable reduction properties) were proved in [379].

Yafaev [534] uses a similar strategy to prove AO for a curve in a Shimura variety (the original conjecture of André), again under GRH, with further results under additional assumptions on the special points. The core result in [534] is a lower bound for the Galois orbits of special points. The unconditional results in [534] improve earlier results along the same lines in [201].

Equidistribution

Two kinds of equidistribution play a role in subsequent work. The possibility of using equidistribution of the Galois orbits of special points (i.e. of suitable analogues of Duke's Theorem [194]) is suggested in [197]. Zhang [548] obtains Galois equidistribtion results for CM-points on quaternionic Shimura varieties under the assumption of suitable subconvexity bounds and bounds on torsion in class groups. The results imply AO unconditionally for certain quaternionic Shimura varieties whose special points satisfy suitable properties.

Equidistribution results for special subvarieties were brought in to the AO context following [140, 141]. See also [149]. The combination of equidistribution results and Galois-theoretic techniques led to a full conditional proof of AO by Klingler, Ullmo, and Yafaev [311, 511].

6.5 Theorem ([311, 511])   *Assuming GRH for CM fields, AO holds in general.*

In these conditional results, GRH is needed for two purposes: to ensure large Galois orbits of special points, and to get an effective Cebotarev density theorem guaranteeing the existence of a small split prime in a number-field. Unconditional results are obtained when restricting to special points in a single Hecke orbit (see Conjecture 18.11). Theorem 6.5 was announced around 2006. More recently, a second proof of AO under GRH, removing ergodic theory and relying instead on degree estimates, has appeared in [162].

Results via Point-Counting

The extension of the Counting Theorem to algebraic points ([424]) and the analogy between MM and AO emphasized in various expositions (e.g. [508, 535]) suggested the idea of applying the point-counting strategy to AO. This resulted initially in a new proof ([425]) of AO in the case of $Y(1)^2$ (and also of some variant problems).

The point-counting strategy is, on its face, applicable to AO in general. For one can always take the Hermitian symmetric domain Ω to be a semi-algebraic open domain, and choose a fundamental domain F for the action of Γ that is semi-algebraic and with a $\overline{\mathbb{Q}}$-structure. Special points in X correspond to suitable algebraic points of bounded degree in Ω, and so on. Whether it succeeds in proving AO for a given Shimura variety depends on the availability of certain key further ingredients. Specifically, Ullmo [509] shows that, given four ingredients, the point-counting strategy indeed proves AO for a given Shimura variety. These ingredients correspond to our Definability, Height, Galois, and Ax–Lindemann ingredients; a property akin to our Finiteness Ingredient then follows.

Building on the work of several people on these ingredients (described further below), a proof of AO for the Siegel modular varieties $\mathcal{A}_g$ is obtained by Tsimerman in [506], and a full proof of AO has been announced in [433].

6.6 Theorem (Tsimerman [506])   *AO holds for $\mathcal{A}_g$ for all $g \geq 1$.*

The final ingredient in the proof of Theorem 6.6, established in [506], is a suitable lower bound for the Galois orbit of a special point. This is deduced, using tools specific to $\mathcal{A}_g$, from a suitable upper bound for the height of a special point. However, Binyamini–Schmidt–Yafaev [79] show that such a height bound implies a Galois orbit lower bound in the general case. The final step is then a suitable height bound for special points, which has been announced by Pila–Shankar–Tsimerman [433].

6.7 Theorem (Pila–Shankar–Tsimerman [433])   *AO holds in general.*

On the generalization of AO to mixed Shimura varieties, see Conjecture 18.8 and Theorem 20.2.

We now give a brief introduction to $\mathcal{A}_g$ and the ingredients for the point-counting strategy leading to the proof of Theorem 6.6. We then indicate the additional steps required for Theorem 6.7, in which point-counting plays a further role.

The Siegel Modular Variety

The following draws on [81, 236] and other sources cited. See also [398]. For a description of $\mathcal{A}_g$ qua Shimura variety, see especially [380].

An *abelian variety* is the higher-dimensional analogue of an elliptic curve. It is, by definition, a complete group variety; that is, a projective variety equipped with a group law (and inverse) given by morphisms. Over $\mathbb{C}$, every abelian variety of dimension g arises as a *complex torus*; that is, a quotient $\Lambda \backslash \mathbb{C}^g$, where $\Lambda \subset \mathbb{C}^g$ is a *lattice*.

However, not every lattice gives rise to an abelian variety; in general, there are no non-constant meromorphic functions on the quotient. A necessary and sufficient condition for $\Lambda \backslash \mathbb{C}^g$ to have the structure of an algebraic variety (and thus to be an abelian variety) is given by the *Riemann relations*, which posit the existence of a positive definite Hermitian form H on $\mathbb{C}^g$ that takes integer values on the lattice. Such an H is called a *polarization*. The polarization has a *degree* $(d_1, \ldots, d_g)$, where the d_i are positive integers with $d_1 | d_2 | \cdots | d_g$. The polarization is *principal* if this sequence is $(1, \ldots, 1)$.

The quasi-projective variety $\mathcal{A}_g$ classifies principally polarized abelian varieties of dimension g over $\mathbb{C}$ up to isomorphism over $\mathbb{C}$. Every elliptic curve over $\mathbb{C}$ is principally polarized, and so $\mathcal{A}_1 = Y(1)$.

As in the case of elliptic curves, lattices in $\mathbb{C}^g$ may be normalized by stipulating some of the basis elements. The space of normalized lattices in $\mathbb{C}^n$ admitting a principal polarization is the *Siegel upper half-space*

$$\mathbb{H}_g = \{Z \in \mathrm{Mat}(g \times g, \mathbb{C}) : Z^t = Z, \mathrm{Im} Z > 0\},$$

where Z^t is the transpose of Z, and $\mathrm{Im} Z > 0$ means that the matrix of imaginary parts is positive definite. The Siegel upper half-space $\mathbb{H}_g$ is a complex space of dimension $g(g + 1)/2$.

The symplectic group

$$\mathrm{Sp}_{2g}(\mathbb{R}) = \{M \in \mathrm{GL}_{2g}(\mathbb{R}) : MJM^t = J\}, \quad \text{where } J = \begin{pmatrix} 0_g & I_g \\ -I_g & 0_g \end{pmatrix}$$

acts on $\mathbb{H}_g$. We have $\dim \mathrm{Sp}_{2g} = 2g^2 + g$. If we write $g \in \mathrm{Sp}_{2g}(\mathbb{R})$ as

$$g = \begin{pmatrix} A & B \\ C & D \end{pmatrix}, \quad A, B, C, D \in \mathrm{Mat}(g \times g, \mathbb{R}),$$

the action is given by

$$gZ = (AZ + B)(CZ + D)^{-1}.$$

(One has $\mathrm{Sp}_2(\mathbb{R}) = \mathrm{SL}_2(\mathbb{R})$.)

Two points in $\mathbb{H}_g$ give rise to isomorphic principally polarized abelian varieties if and only if they are in the same $\mathrm{Sp}_{2g}(\mathbb{Z})$ orbit. Thus, the quotient $\mathrm{Sp}_{2g}(\mathbb{Z}) \backslash \mathbb{H}_g$ classifies them. It has the structure of a quasi-projective variety, and is denoted by $\mathcal{A}_g$. The uniformization

$$u_g \colon \mathbb{H}_g \to \mathcal{A}_g$$

may be given by suitable theta functions ([384]). The variety $\mathcal{A}_g$, like $\mathcal{A}_1$, is not complete.

The special subvarieties of $\mathcal{A}_g$ are not easy to describe explicitly. They correspond to spaces of abelian varieties with additional structure of various kinds, including (but not limited to) additional endomorphisms, and/or level structure. They are variously characterized in [380], and the following characterization is given in [395].

6.8 Characterization ([395]) A *special subvariety* of $\mathcal{A}_g$ can be characterized as a Zariski-closed subvariety $T \subset \mathcal{A}_g$ with the property that there exists an algebraic subgroup $H \subset \mathrm{Sp}_{2g}$ defined over $\mathbb{Q}$ and a special point $Z \in \mathbb{H}_g$ (see Characterization 6.9) such that T is the image of $H(\mathbb{R})^c \cdot Z$ in $\mathcal{A}_g$.

Here $H(\mathbb{R})^c$ is the connected component (in the real topology) of the identity. In general, however, the image of $H(\mathbb{R})^c \cdot Z$ in $\mathcal{A}_g$ is not Zariski closed ([395]). A special subvariety is associated with a morphism of Shimura data ([163, 17.1]), and one can assume that the subgroup H in Characterization 6.8 is semi-simple and defined over $\mathbb{Q}$ ([435, 2.3]). There are only finitely many semi-simple real groups that embed into $\mathrm{Sp}_{2g,\mathbb{R}}$ and the embeddings come in finitely many families up to conjugacy (over $\mathbb{R}$); see [202]. The special subvarieties of $\mathcal{A}_g$ include certain Shimura curves and Hilbert modular varieties. Also $Y(1)^g$ appears as a special subvariety of $\mathcal{A}_g$. An explicit classification of the special subvarieties of $\mathcal{A}_2$ is given in [167].

6.9 Characterization   The *special points* of $\mathcal{A}_g$ are the points corresponding to CM abelian varieties (see below). A *special point* of $\mathbb{H}_g$ is any pre-image of a special point in $\mathcal{A}_g$.

As noted in [395], one must be careful in describing when an abelian variety A is CM. It is common to say that *"A has CM by F"* to mean that F embeds into the *endomorphism algebra*

$$\mathrm{End}^0 A = \mathrm{End}(A) \otimes_{\mathbb{Z}} \mathbb{Q}$$

of A. Here $\mathrm{End}(A)$ is the *endomorphism ring* of A, consisting of endomorphisms defined over some fixed algebraic closure of its field of definition (always defined over an extension of the field of moduli of degree at most $2 \cdot 3^{4g^2}$; see [506, Lemma 4.1]). Generically, one has $\mathrm{End}(A) = \mathbb{Z}$. For an elliptic curve and an imaginary quadratic field this indeed suffices to characterize CM. However, to be CM requires, in general, as many as possible complex multiplications, and this requires some care in formulation for higher-dimensional abelian varieties.

An *isogeny* between abelian varieties is a surjective homomorphism with finite kernel. If an abelian variety A has a non-trivial proper abelian subvariety B, then it is isogenous to a product $B \times C$ for a suitable complementary abelian subvariety (*Poincaré reducibility*). An abelian variety with no non-trivial proper abelian subvarieties is called *simple*. An abelian variety is thus isogenous to a product of simple abelian varieties. A simple abelian variety A is CM if $\mathrm{End}^0 A$ contains a subfield of degree $2 \dim A$ over $\mathbb{Q}$ (over a field of characteristic 0, this is as large as possible). A general, not necessarily simple, abelian variety A is CM if there is an isogeny of A with $\sum_i B_i$, where B_i are simple and CM. Over $\mathbb{C}$, A has CM if and only if its Mumford–Tate group is commutative ([199]).

Note that lying on a proper special subvariety does not always entail having extra endomorphisms, as shown by the construction in [382] of a special curve in $\mathcal{A}_4$ that is not of *PEL type*. These Mumford–Shimura curves are studied in [390]. The corresponding abelian varieties have, generically, endomorphism ring equal to $\mathbb{Z}$.

Ingredients for AO via Point-Counting

We consider the uniformization

$$u \colon \Omega \to X$$

of a Shimura variety, invariant under the action of Γ. We discuss the ingredients as in the proof of the modular case, though out of order to give the Galois orbit bounds their place as the last step.

6.10 Definability Ingredient

One needs definability in an o-minimal structure of the restriction of $u \colon \Omega \to X$ to a fundamental domain (or fundamental set) for the Γ-action to serve as a definable set for point-counting. In the co-compact case, when X is projective, one immediately gets definability (in the o-minimal structure $\mathbb{R}_{an}$; see §8.21). But it is non-trivial to establish definability in the presence of cusps.

In the modular case, the definability of j restricted to the classical fundamental domain was observed by Peterzil–Starchenko [411], as a consequence of its q-expansion, in the course of proving suitable definability for the Weierstrass $\wp$ function as a function of two variables. The generalization to $\mathcal{A}_g$ is proved by them in [415]. Klingler–Ullmo–Yafaev [309] prove the result for the uniformization $u \colon \Omega \to \Gamma \backslash \Omega$ of any arithmetic variety (i.e. Ω a Hermitian symmetric domain and Γ an arithmetic subgroup of its group of bi-holomorphisms). So it holds for every (mixed) Shimura variety.

6.11 Theorem ([309])   *Let $u \colon \Omega \to X$ be the uniformizing map of an arithmetic variety. Then there exists a semi-algebraic fundamental set F for the action of Γ on Ω such that the restriction $u|_F \colon F \to X$ is definable in $\mathbb{R}_{an,\,exp}$.*

Bakker–Klingler–Tsimerman [29] prove that the period map associated to any pure polarized variation of Hodge structure on a smooth complex quasi-projective variety is definable. They use this and o-minimal GAGA (see Chapter 12) to obtain a new proof of the fundamental theorem of Cattani–Deligne–Kaplan on the Hodge locus. This is extended to mixed period maps in [26].

Definability results for Riemann mappings of semi-analytic plane domains are established in [301].

6.12 Height Ingredient

Taking Ω with its $\overline{\mathbb{Q}}$-algebraic structure, the pre-image of a special point $x \in X$ is algebraic and of some bounded degree depending only on X (see the discussion in [166, §1.2]; this follows for example from [164, Theorem 2.3]).

We need a suitable bound for the height of a pre-image $\omega \in F$ of a special point $x \in X$ in terms of a complexity measure for x. In the case of $\mathcal{A}_g$, a natural choice of complexity measure is the size of the discriminant of the centre

$$R_x = Z\big(\mathrm{End}(A_x)\big)$$

of the endomorphism ring of the corresponding abelian variety A_x. We set

$$\Delta(x) = |\mathrm{Disc}(R_x)|.$$

For the general case one defines the complexity in terms of the Hodge structure associated with x, see [166] (also [396]).

6.13 Theorem ([166, Theorem 1.1]) *With suitable choices of the realization of Ω and the fundamental domain F, there exist constants c, γ (depending on X) such that*

$$H(\omega) \le c\Delta(x)^{\gamma}.$$

For the modular case this is elementary (Ingredient 5.4). Going further it becomes a non-trivial problem in reduction theory. For $\mathcal{A}_g$ it is due to Tsimerman, and was published in [434]. It is proved for general Shimura varieties by Daw–Orr [166].

6.14 Ax–Lindemann Ingredient

The key functional transcendence statement ensuring that algebraic subvarieties of $u^{-1}(V)$ are accounted for by weakly special subvarieties.

6.15 Theorem (Ax–Lindemann for Shimura varieties [309, 427, 434, 435, 514]) *A maximal algebraic $W \subset u^{-1}(V)$ is weakly special.*

The result is extended to the mixed case in [225, 226].

Ax–Lindemann for a Shimura variety is a special case of Ax–Schanuel [377], which is the analogue for u of the classical Ax–Schanuel Theorem for the exponential function. We return to this topic in Part III.

Ax proved his theorem in the setting of differential algebra, giving a second proof via differential geometry ([21]), and generalizing it to certain commutative algebraic groups, including semi-abelian varieties (see [50]). A proof of Ax–Lindemann for abelian varieties using value distribution theory is given in [389], and one for semi-abelian varieties using only basic properties of o-minimal structures is given in [416].

For the uniformizing maps of Shimura varieties, Ax–Lindemann and Ax–Schanuel are proved in the complex setting combining point-counting with ideas from complex geometry (in particular volume estimates of Hwang–To [290]), monodromy (especially the theorem of Deligne–André [5, 172, 173]), and tame complex analysis due to Peterzil–Starchenko [409, 410, 414] (see also §12). A proof of Ax–Lindemann for ball quotients using methods of several complex variables is in [376].

A new proof of Ax–Lindemann for the modular function (with derivatives, as in [428]), and more generally for the uniformizing maps of genus zero Fuchsian groups, is given in [121] via differential algebra, in particular Picard–Vessiot theory, monodromy, and model theory. This approach is generalized further in [82, 83]. Ax–Lindemann results for quasi-modular functions and for almost holomorphic modular functions are established in [493].

A non-archimedean Ax–Lindemann Theorem for the Schottky uniformization of products of hyperbolic Mumford curves is in [129], using the non-archimedean point-counting results in [142]. (See also [289].)

6.16 Galois Ingredient

Specifically, one requires a bound of the form

$$\#\left(\mathrm{Gal}(\overline{\mathbb{Q}}/\mathbb{Q}) \cdot x\right) \geq c\Delta(x)^{\delta}$$

for the Galois orbit of a special point $x \in X$, for some positive constants c, δ (depending on X). In the basic case of $Y(1)$, this is afforded (ineffectively) by Siegel's Theorem ([491]).

The question of such a bound was raised by Edixhoven [199, Problem 14], observing its utility for AO. Somewhat weaker bounds were established in [511, 534] under GRH. Unconditional bounds were established in [513] (for $g \leq 3$) and [504] (for $g \leq 6$). Both establish such bounds for all g under GRH, and unconditionally for *Weyl CM points*.

Tsimerman [506] establishes a bound of the desired form for $\mathcal{A}_g$, completing the proof of AO for $\mathcal{A}_g$.

6.17 Theorem ([506, Theorem 5.2]) *Let $g \geq 1$. There exist constants c_g, δ_g such that, for a special point $x \in \mathcal{A}_g$, one has*

$$\#\left(\mathrm{Gal}(\overline{\mathbb{Q}}/\mathbb{Q}) \cdot x\right) \geq c_g\Delta(x)^{\delta_g}.$$

The Galois lower bound is deduced from a suitable upper bound for the *Faltings height* of the corresponding abelian variety (see e.g. [408]) using the isogeny estimates of Masser–Wüstholz [355, 356]. Specifically, one needs a bound of the form

$$h_{\mathrm{Fal}}(A_x) \ll_{\epsilon} \Delta(x)^{\epsilon}$$

(for fixed g) when x is a CM point of $\mathcal{A}_g$ with primitive CM type. A conjecture of Colmez [152] predicts a formula for the Faltings height in terms of special values of L-functions. This conjecture remains open in general, but Tsimerman shows that the required height bound can be deduced from the averaged Colmez conjecture. This had just been proved at the time, independently and by different methods, by Andreatta–Goren–Howard–Madapusi Pera [11] and Yuan–Zhang

[540]. The Faltings height and the height of the corresponding moduli point are close (the difference is logarithmically bounded: see e.g. [408]).

The proof of the Galois lower bound required for Theorem 6.6 thus utilizes two inputs specific to the $\mathcal{A}_g$ setting: the Faltings height of an abelian variety (and the averaged Colmez conjecture for it), and the Masser–Wüstholz estimates for isogenies between abelian varieties. Neither of these is available for a general Shimura variety, which need not parameterize abelian varieties.

We now briefly indicate the additional ingredients required to prove Theorem 6.7. First, a sufficiently strong point-counting result (in terms of its dependence on both the height and degree of the points being counted) can be used to replace isogeny estimates for abelian varieties, and deducee Galois lower bounds in the presence of height upper bounds.

This idea is due to Schmidt [475], who implements it for torsion points in group varieties in basic cases. Point-counting bounds of the required strength are not available in general, but Binyamini [71] establishes counting results of the right quality for foliations defined over number-fields, which are applicable for general Shimura varieties. This approach is implemented by Binyamini–Schmidt–Yafaev [79], who show that a suitable height upper bound for special points implies the Galois lower bounds. See the further discussion in Chapter 11.

The height estimate is supplied in [433]. This relies on an idea of Deligne [174] for combining suitable partial CM types into a complete CM type, and on the construction of an (almost) canonical height on a general Shimura variety. Thus, point-counting is employed at three separate junctures in the proof of the André–Oort conjecture. Only this final application to obtain Galois orbit lower bounds requires a strong quantitative point-counting result.

The role of height upper bounds here reprises a similar role in unlikely intersection results in group varieties, starting with the work of Bombieri–Masser–Zannier [87, 543] in the multiplicative setting and their Bounded Height conjecture [90]. This was affirmed in [252] (see Theorem 20.7 and, in the abelian setting, before Theorem 20.10). In contrast, the failure of bounded height for special points of $Y(1)$ is observed in [90], using results on the size of coefficients of modular polynomials ([147]). A Weakly Bounded Height conjecture in the modular setting is formulated in [254], which, as shown in [264] (and see Chapter 21), would suffice even in a weaker form to imply the full Zilber–Pink conjecture for a product of modular curves.

Effective and Explicit Results

The unconditional proof of AO for $Y(1)^2$ by André is ineffective, as are all proofs using the basic point-counting strategy. The source of ineffectivity in

André's proof, and one source of it in the point-counting proofs, is the reliance on positive power of discriminant lower bounds for class numbers. All known bounds of this quality depend on Landau–Siegel: the known effective bounds (see e.g. [246]) are logarithmic.

In the point-counting proofs there is a second source of ineffectivity. Namely, the upper bounds from point-counting are in general not effective. Unlike the case of Landau–Siegel, where effectivity would seem to require a proof of GRH, point-counting in these settings is presumably not inherently ineffective, but getting effectivity seems to be difficult. See Chapter 10 for further discussion. A third ineffective step is the finiteness property in 5.13 and its analogues. In the modular case, this is made effective in [70]; in general, see [73, 165].

To our knowledge, the only Shimura variety (of dimension ≥ 2) for which AO is known effectively is $Y(1)^2$, due independently to Kühne [317] (see also [318]) and Bilu–Masser–Zannier [67]. See also the variant argument in [532] and the extension of effectivity to special points on curves in powers of elliptic modular surfaces. These proofs use linear forms in logarithms (classical and elliptic). (The proof in [196] of AO for $Y(1)^2$ under GRH is made effective under GRH in [101].)

In the following, $h(V)$ is the projective height of a polynomial $P \in \overline{\mathbb{Q}}[X_1, X_2]$ defining the curve V.

6.18 Theorem ([317, Theorem 2]) *For all $\epsilon > 0$ and positive integers δ, D, there exists an effectively computable constant $C(\epsilon, \delta, D) > 0$ with the following property. Let $V \subset Y(1)^2$ be a geometrically irreducible algebraic curve defined over a number-field K. For $i = 1, 2$, let δ_i denote the degree of $X_i|_V \to \mathbb{C}$ and assume $\delta_i > 0$. Then,*

$$\max\{|\Delta_1|, |\Delta_2|\} < C(\epsilon, \max\{\delta_1, \delta_2\}, [K : \mathbb{Q}]) \max\{1, h(V)\}^{8+\epsilon}$$

for every CM point of discriminant (Δ_1, Δ_2) on V that does not lie on any modular curve T_N with $1 \leq N \leq 4 \max\{\delta_1, \delta_2\}^5$.

The proofs in [67, 317] can be carried out explicitly in certain cases to show that, for example, there are no solutions in singular moduli to the equations $x + y = 1$ ([318]) and $xy = 1$ ([67]). AO is proved effectively for all linear subvarieties of $Y(1)^n$ in [64], using class field theory ([320]). These ideas have enabled effective (sometimes explicit) resolution of equations in singular moduli of specific forms ([1, 66, 216, 218, 339, 460]).

Bilu–Luca–Masser [65] establish a finiteness result for collinear triples of CM points in $Y(1)^2$. By studying the possible linear dependencies of functions of the form $j(g_i z)$, where $g_i \in \mathrm{GL}_2^+(\mathbb{Q})$, they show that the corresponding incidence locus has no non-trivial positive-dimensional special subvarieties.

This leaves a finite number of possible examples, but these are not effectively controlled. Multiplicative dependencies among singular moduli are studied in [437]. See further [217].

Binyamini [70] employs Siegel–Tatuzawa (essentially, that there is at most one bad quadratic field obstructing an effective Siegel bound) and Duke's Theorem [194], together with an effective Ax–Lindemann result obtained using bounds from differential algebra and effective point-counting results for semi-Noetherian sets ([69]) (see Definition 10.17) to get an almost fully effective result for subvarieties of $Y(1)^n$. Essentially, the point-counting is made effective on compact domains, an effective equidistribution is obtained excluding the bad field, and Galois conjugates are taken over class fields corresponding to singular moduli from the bad field; this yields a fully effective result if one excludes points all of whose coordinates come from the one bad quadratic field. The point-counting is improved in [71] to poly-logarithmic on compact sets with a polynomial dependence on the inverse distance to the anomalous locus. This leads to an ineffective polynomial bound on the special set in terms of the logarithmic height of $V \subset Y(1)^n$, its degree, and the degree of its field of definition. See also [71].

Under mild conditions, Binyamini–Masser [74] establish AO effectively for non-compact curves in Hilbert modular varieties of dimension $g = 2$ or odd. The proof is via point-counting, gaining effectivity in this aspect all the way to the cusp through the theory of *Q-functions* and results of [4] on G-*functions* (see also [167]). The non-compactness is essential for the Galois orbit bounds.

It is shown in [63] that no singular modulus is an algebraic unit, improving an earlier ineffective finiteness result in [259]. As a consequence, there can be no special points on any subvariety of $Y(1)^n$ defined, for example, by $x_1 P(x_1, \ldots, x_n) = 1$, where $P \neq 0$ is a polynomial with algebraic integer coefficients. The results of [63] are generalized in [334] using other methods. On singular moduli that are S-units, see [116, 276].

Generalizations, Variants, and Analogues

André [7] formulated a Generalized André–Oort conjecture for mixed Shimura varieties. See Conjecture 18.8 and Theorem 20.2. Klingler [308] generalizes AO (and more generally the Zilber–Pink conjecture) to the setting of variations of Hodge structure. See [136] for geometric results. See also [33]. Variants of AO for certain non-classical modular functions are studied in [493, 494, 495].

A p-adic version of AO for the universal family of abelian varieties is given in [468], via the model theory of difference fields. A mod p variant is formulated and studied (under GRH) in [200], and a two-dimensional arithmetic version is framed in [457]. Results on an analogue of AO for Drinfeld modular varieties are obtained in [102, 285]. A dynamical analogue of André–Oort is formulated in [25]; see also [239].

AO is connected to Coleman's conjecture ([557, Conjecture 6]) on the finiteness of the number of Jacobians with CM for a given (sufficiently large) genus; see, for example, [380].

PART II

O-MINIMALITY AND POINT-COUNTING

7

Model Theory and Definable Sets

We now turn to the notion of a set being definable in an o-minimal structure. To contextualize this notion, we need to go back to Tarski and his proof ([499]) of the decidability of the theory of real closed fields. And to do that we need to introduce the basic setting of model theory, which we want also in connection with the provenance of the Zilber–Pink conjecture in Chapter 18. A standard introductory reference on model theory is [345].

The Setting of Model Theory

In model theory one studies systems of statements in a first-order language, and the mathematical structures that satisfy them. First order means that quantification $\forall$ and $\exists$ runs only over the universe of the structure, and not, for example, over its subsets. That would be second order. Many of the usual structures in mathematics can be axiomatized in a first-order setting.

7.1 Example (Fields)   A *field* consists of a set F together with some additional structure: the binary functions giving addition and multiplication, and their respective identity elements. In model theory one sets up a corresponding *language*; in this case the *language of rings*

$$\mathcal{L}_{\text{rings}} = \{+, \times, 0, 1\},$$

where $+, \times$ are *(binary) function symbols* and $0, 1$ are *constant symbols*. By convention one always includes $=$ as well as the exhibited symbols of a language, along with the usual logical connectives $\wedge$ (conjunction), $\vee$ (disjunction), $\neg$ (negation), $\rightarrow$ (implication) and $\leftrightarrow$ (iff), and the quantifiers $\forall, \exists$. The language also includes an infinite supply of variables, for which we use x, y, z, and so on, with or without subscripts.

We next define an $\mathcal{L}_{\text{rings}}$-*structure*, which is a tuple of the form

$$\mathcal{F} = (F, +_{\mathcal{F}}, \times_{\mathcal{F}}, 0_{\mathcal{F}}, 1_{\mathcal{F}}).$$

This consists of a set F equipped with *interpretations* of $+, \times$ as binary functions $+_{\mathcal{F}}, \times_{\mathcal{F}} \colon F \times F \to F$, and *interpretations* $0_{\mathcal{F}}, 1_{\mathcal{F}} \in F$ of $0, 1$. Usually one abuses the notation by writing simply

$$\mathcal{F} = (F, +, \times, 0, 1),$$

neglecting to distinguish between the symbols of the language and their interpretation in a structure (even though this distinction is the crux of model theory).

An $\mathcal{L}_{\text{rings}}$-structure need not however be a field, or even a ring. An $\mathcal{L}$-structure $\mathcal{F} = (F, +, \times, 0, 1)$ is a field if it satisfies certain *axioms*. This means that various *sentences* of the language; that is, formulae without free variables, hold true in the structure under the given interpretations of the symbols, and where quantifiers range over the underlying set F. Axioms for a field include associativity of addition, existence of multiplicative inverses, and so forth:

$$\forall x \forall y \forall z (x + y) + z = x + (y + z),$$
$$\forall x (\neg x = 0 \to \exists y \, xy = 1), \quad \text{etc.}$$

The point is that all the axioms are expressible as first-order statements in $\mathcal{L}_{\text{rings}}$.

More generally, one defines first-order languages and their corresponding structures as follows.

7.2 Definition   A *first-order language* $\mathcal{L}$ consists of an arbitrary number of constant symbols, predicate (i.e. relation) symbols and function symbols, of various arities,

$$\mathcal{L} = \{c_i, i \in I; , R_j, j \in J; f_k : k \in K\},$$

together with the logical symbols $=, \wedge, \vee, \neg, \to, \leftrightarrow$, as well as variables as above and brackets for readability.

7.3 Definition   Given a first-order language

$$\mathcal{L} = \{c_i, i \in I; , R_j, j \in J; f_k : k \in K\},$$

one has a corresponding notion of an $\mathcal{L}$-*structure*

$$\mathcal{A} = (A, c_i^{\mathcal{A}}, i \in I, R_j^{\mathcal{A}}, j \in J, f_k^{\mathcal{A}}, k \in K),$$

in which A is a set, $c_i^{\mathcal{A}}$ are elements of A, $R_j^{\mathcal{A}}$ are predicates on (i.e. subsets of) the appropriate cartesian power of A^{r_j}, where r_j is the arity of R_j, and $f_k^{\mathcal{A}} \colon A^{s_k} \to A$, where s_k is the arity of f_k. Again, and as is traditional, we do not observe the distinction between the symbols of the language and their interpretation in a particular structure, and we drop the "$\mathcal{A}$".

One next has a notion of $\mathcal{L}$-*formula*. These are the meaningful strings in the language, though the definition of them is purely syntactic.

7.4 Definition Let $\mathcal{L}$ be a first-order language.

1. The set of *terms* of $\mathcal{L}$ is defined inductively as follows:
   (i) A constant symbol is a term; a variable is a term.
   (ii) If $t_1, \ldots, t_n$ are terms and f an n-ary function symbol, then $f(t_1, \ldots, t_n)$ is a term.
2. The set of *atomic formulae* of $\mathcal{L}$ consists of:
   (i) $s = t$, where s, t are terms.
   (ii) $R(t_1, \ldots, t_n)$, where $t_1, \ldots, t_n$ are terms and R is an n-ary relation symbol.
3. The set of *formulae* of $\mathcal{L}$ is defined inductively by:
   (i) An atomic formula is a formula.
   (ii) If ϕ, ψ are formulae, then $\neg\phi$, $(\phi \wedge \psi)$, $(\phi \vee \psi)$, $(\phi \to \psi)$, and $(\phi \leftrightarrow \psi)$ are formulae.
   (iii) If ϕ is a formula and x is a variable, then $\exists x\phi$ and $\forall x\phi$ are formulae.

7.5 Definition 1. An occurrence of a variable in a formula is called *bound* if it is in the scope of a quantifier. If it is not bound, then the occurrence is *free*.
2. A formula with no free occurrences of variables is called a *sentence*.

7.6 Definition Let $\mathcal{L}$ be a first-order language, and $\mathcal{A}$ an $\mathcal{L}$-structure.

1. An $\mathcal{L}$-*theory* is a set T of $\mathcal{L}$-sentences.
2. Let ϕ be an $\mathcal{L}$-sentence. The interpretation of ϕ may or not hold true in $\mathcal{A}$. If it does, we say that $\mathcal{A}$ *models* ϕ and write $\mathcal{A} \models \phi$.
3. More generally, if $\phi(x_1, \ldots, x_n)$ is an $\mathcal{L}$-formula with free variables among $x_1, \ldots, x_n$, and $a_1, \ldots, a_n \in A$, then we write

$$\mathcal{A} \models \phi(a_1, \ldots, a_n)$$

if $\phi(x_1, \ldots, x_n)$ holds in $\mathcal{A}$ with the indicated substitution. (This notion requires a little care to formulate properly as one does not want to substitute bound variables.)
4. Let T be an $\mathcal{L}$-theory. If $\mathcal{A} \models \phi$ for every $\phi \in T$ we say that $\mathcal{A}$ is a *model* of T and write $\mathcal{A} \models T$.
5. The set of all $\mathcal{L}$-sentences that are true in $\mathcal{A}$ is a theory called *the theory of* $\mathcal{A}$, and denoted as $\text{Th}(\mathcal{A})$.

6. Two $\mathcal{L}$-structures $\mathcal{A}, \mathcal{B}$ with the same theory are called *elementarily equivalent* and one writes $\mathcal{A} \equiv \mathcal{B}$.

7. A theory T has *quantifier elimination* if, for every formula ϕ, there is a quantifier-free formula ψ such that $T \models \phi \leftrightarrow \psi$.

8. A theory T is *model complete* if (equivalently), for every formula ϕ, there is an existential formula ψ such that $T \models \phi \leftrightarrow \psi$.

A language $\mathcal{L}$ may be expanded by further symbols to a richer language $\mathcal{L}'$. Correspondingly, an $\mathcal{L}$-structure $\mathcal{A}$ may be expanded to an $\mathcal{L}'$-structure $\mathcal{A}'$ by choosing interpretations for the new symbols.

7.7 Definition Suppose $\mathcal{L} \subset \mathcal{L}'$ are first-order languages, and $\mathcal{A}$ is an $\mathcal{L}$-structure.

1. An $\mathcal{L}'$-structure, structure $\mathcal{A}'$, formed from $\mathcal{A}$ by choosing interpretations for the symbols in $\mathcal{L}' \backslash \mathcal{L}$ is called an *expansion* of $\mathcal{A}$. If one wants to indicate such an expansion without specifying precisely the additional structure involved, one writes

$$\mathcal{A}' = (\mathcal{A}, \ldots),$$

where the "dots" then indicate some unspecified additional constants, predicates, and/or function symbols. In particular, a structure $\mathcal{A}$ on a set A can be generically written as

$$\mathcal{A} = (A, \ldots).$$

2. Conversely, if $\mathcal{A}'$ is an $\mathcal{L}'$-structure, then forgetting the symbols in $\mathcal{L}' \backslash \mathcal{L}$ gives an $\mathcal{L}$-structure $\mathcal{A}$ called a *reduct* of $\mathcal{A}'$.

Completeness and Compactness

We do not need to dwell on the syntactic side, but associated with a first-order language is a notion of a *derivation*, in which a formula ϕ is derived from a set of formulas T by some formal rules. One then writes $T \vdash \phi$. A derivation has (by definition) only finitely many steps.

Suppose that $\mathcal{L}$ is a first-order language and T is an $\mathcal{L}$-theory. We now have two notions of when a sentence ϕ follows from T. One is $T \vdash \phi$. The other, the notion of *logical consequence*, is that ϕ holds in any $\mathcal{L}$-structure in which all the sentences in T hold, denoted as $T \models \phi$.

The *Completeness Theorem* (for first-order predicate calculus), proved by Gödel, asserts that these two notions coincide. This is the foundational result for model theory.

A key consequence is the *Compactness Theorem*: a theory is *consistent* (i.e. has a model) if and only if all of its finite subsets are consistent.

Proof: the derivation of a contradiction would involve only finitely many of the axioms from *T*.

Categoricity

Compactness asserts that any finitely realizable set of sentences has a model, and leads to the construction of non-standard models. It also means that if a theory *T* has an infinite model, then it has models of arbitrarily large cardinality. *Proof:* Suppose *T* has a model $\mathcal{M}$ of cardinality μ. For any cardinal $\kappa > \mu$, expand the language by a set of new constant symbols of size κ and introduce new axioms asserting that their realizations are pairwise distinct. This theory is consistent by compactness, and so has a model. If we take $T = \text{Th}(\mathcal{M})$, then the new model, of size at least κ, is elementarily equivalent to $\mathcal{M}$.

Thus, a theory, provided it has infinite models, can never determine its models up to isomorphism. The best one can get (in terms of a theory determining as much as possible about its models as structures of the relevant language) is that it determines them up to isomorphism in a given cardinality.

7.8 Definition Let *T* be a theory and μ a cardinal number. Then, *T* is said to be *categorical in cardinality* μ if all models of *T* of size μ are *isomorphic* (as $\mathcal{L}$-structures; that is, the isomorphism is a bijective map under which interpretations of the constants, functions, and relations of $\mathcal{L}$ correspond).

This is a very strong property in view of the fact that any finitely consistent set of sentences containing that theory can be realized in some model of it.

7.9 Example (Algebraically closed fields) Clearly, not all fields are elementarily equivalent, as while some fields contain a solution to $x^2 = 2$, others do not, some have characteristic 5 and others do not, and so on. However, algebraically closed fields of a given characteristic are all elementarily equivalent ([346, Corollary 3.2.3]).

Algebraically closed fields can be axiomatized in the first-order setting, though it requires infinitely many axioms (one in each degree, asserting that any polynomial of that degree has a root, also specifying that characteristic zero requires infinitely many axioms, one for each prime, asserting $0 \neq 1 + \cdots + 1$ (with *p* summands).

Consider the theory ACF_0 of algebraically closed fields of characteristic zero. A standard result of field theory is that algebraically closed fields of the same transcendence degree (i.e. the cardinality of a transcendence basis) are isomorphic. Thus, ACF_0 is categorical in all uncountable powers. Therefore, $\mathbb{C}$ is, up to isomorphism, the unique algebraically closed field of characteristic zero and continuum size, a fact enabling properties in field theory to be deduced by methods of complex analysis (the *Lefschetz principle*).

Some Further Examples

7.10 Example (Dense linear order without endpoints)

A simple example is the language $\{<\}$ consisting of a single binary relation symbol. A $\{<\}$-structure is then a pair $\mathscr{A} = (A, <)$, where A is a set and $<$ is a binary relation on A. So far no particular properties are required of $<$, but in this language one can formulate the axioms that $<$ is a strict total order: for example, transitivity corresponds to

$$\forall x \forall y \forall z \big((x < y \wedge y < z) \to x < z\big),$$

and so on, and further that the order is dense and without endpoints:

$$\forall x \forall y \exists u \exists v \exists w \big(u < x \wedge x < v \wedge v < y \wedge y < z\big).$$

Together these give a *theory* known as Dense Linear Order without endpoints (DLO). Examples of dense linear orders without endpoints include

$$(\mathbb{Q}, <), \quad (\mathbb{R}, <), \quad (\mathbb{R} \cap \overline{\mathbb{Q}}, <), \quad ((0,1), <), \quad \text{and so on.}$$

Dense Linear Order is $\aleph_0$-categorical but is not categorical in any uncountable power (**exercise**).

7.11 Example (Exponential fields)

Here one considers a field (or ring) together with a homomorphism from F to $F^\times$, so as a structure it is just the ring structure augmented by an additional unary function. In the corresponding language one can impose that the map is a homomorphism, that it is surjective (or onto some particular subset like $\mathbb{R}_{>0}$), and so on. See [551], for example. Particularly interesting classical structures of this type are the *real exponential field*

$$\mathbb{R}_{\exp} = (\mathbb{R}, <, +, \times, 0, 1, e^x),$$

where the language has been augmented also by the order relation $<$, and the *complex exponential field*

$$\mathbb{C}_{\exp} = (\mathbb{C}, +, \times, 0, 1, e^z).$$

The first is the structure whose study motivated the idea of o-minimality. The second is the setting for Zilber's work on pseudo-exponentiation and the context in which he formulated the (semi-abelian) Zilber–Pink conjecture.

Definable Sets

7.12 Definition   Let $\mathcal{L}$ be a first-order language and $\mathcal{A} = (A, \ldots)$ an $\mathcal{L}$-structure.

1. A *definable set* in $\mathcal{A}$ is a set $Z \subset A^n, n = 1, 2, 3, \ldots$ such that, for some $\mathcal{L}$-formula $\phi(x_1, \ldots, x_n)$, one has

$$Z = \{(x_1, \ldots, x_n) \in A^n : \mathcal{A} \models \phi(x_1, \ldots, x_n)\},$$

where it is assumed that the free variables in ϕ are a subset of $\{x_1, \ldots, x_n\}$.

2. A *definable set with parameters* is a definable set in which the formula ϕ is permitted to contain names for some elements of A. Formally, one has a formula $\phi(x_1, \ldots, x_n, y_1, \ldots, y_m)$ and $a_1, \ldots, a_m \in A$ such that

$$Z = \{(x_1, \ldots, x_n) \in A^n : \mathcal{A} \models \phi(x_1, \ldots, x_n, a_1, \ldots, a_m)\}.$$

7.13 Examples

1. The theory of the complex field $(\mathbb{C}, +, \times, 0, 1)$ has quantifier elimination. This is a consequence of Chevalley's Theorem that the projection of a constructible set is constructible (which follows from elimination theory). Hence, the definable sets with parameters are exactly the *constructible* sets. The definable sets without parameters are the constructible sets defined over $\mathbb{Q}$.

2. The theory of the real field also has quantifier elimination (Tarski–Seidenberg Theorem; see §8), so the definable sets with parameters are precisely the semi-algebraic sets (see e.g. [189, 2.11]). As every real algebraic number is definable using its minimal polynomial and a suitable rational interval in which it is the only root, the definable sets without parameters are the semi-algebraic sets defined over $\mathbb{R} \cap \overline{\mathbb{Q}}$. Also, one can define the relation $<$ by

$$x < y \quad \Longleftrightarrow \quad \{(x, y) \in \mathbb{R}^2 : \exists z : z \neq 0 \wedge y = x + z^2\}.$$

So the definable sets in $(\mathbb{R}, <, +, \times, 0, 1, \ldots)$ and in $(\mathbb{R}, +, \times, 0, 1, \ldots)$ are the same.

3. In a structure for the language of ordered fields, one can define, for example, the closure and interior of a definable set in the order topology, the set of points where a definable function is continuous, differentiable, and so on (**exercises**).

4. In the complex exponential field $\mathbb{C}_{\exp}$, one can define the integers

$$\mathbb{Z} = \{z \in \mathbb{C} : \forall x \ \exp(x) = 1 \rightarrow \exp(zx) = 1\}.$$

A conjecture (attributed to Koiran) is that the set of real numbers is not definable there (even in the expansion of the complex field by *all* univariate entire functions).

5. In $(\mathbb{Z}, +, \times, 0, 1)$, one can define $\mathbb{N}$, using Lagrange's Theorem. In $(\mathbb{Q}, +, \times, 0, 1)$, one can define $\mathbb{Z}$ ([461]).

Minimality and Strong Minimality

Another strong property enjoyed by ACF_0 is *strong minimality*.

7.14 Definition (See e.g. [346, 6.1.2]) A structure $\mathcal{A} = (A, \ldots)$ is called *minimal* if every definable subset of A is either finite or cofinite. It is *strongly minimal* if every structure elementarily equivalent to $\mathcal{A}$ is minimal.

An example of a structure that is minimal but not strongly minimal is given in [346, p. 208].

Minimality fails in general for ordered structures, in which an interval and its complement are typically both infinite. In this context, o-minimality captures the notion of an ordered structure in which the collection of definable subsets of the underlying set (though not of its cartesian powers) is as small as possible.

Minimality also fails in $\mathbb{C}_{\exp}$, as $\mathbb{Z}$ and $\mathbb{C} - \mathbb{Z}$ are both infinite. However, it is conjectured (and in a stronger form) by Zilber [549, 551] that $\mathbb{C}_{\exp}$ is *quasi-minimal*, namely, that every definable subset of $\mathbb{C}$ is either countable or co-countable. For now it is unknown whether $\mathbb{R} \subset \mathbb{C}$ is a definable subset in $\mathbb{C}_{\exp}$ (Koiran's conjecture; see Example 7.13.4). For affirmation in a special case, see [96].

8

O-Minimal Structures

A Question of Tarski

The circle of ideas around Gödel's Incompleteness Theorem shows that the basic theory of the natural numbers (Peano Arithmetic) is undecidable. It follows, using the definability of $\mathbb{N}$ in $\mathbb{Z}$, that the $\mathcal{L}_{\text{rings}}$-theory of the integers $(\mathbb{Z}, +, \times, 0, 1)$ is undecidable and (using the definability of $\mathbb{Z}$ in $\mathbb{Q}$; see Example 7.13.5) that $\text{Th}(\mathbb{Q}, +, \times, 0, 1)$ in $\mathcal{L}_{\text{rings}}$ is undecidable.

However, Tarski [499] proved that the $\mathcal{L}_{\text{rings}}$-theory of the real field is decidable. This theory is just the theory of real closed fields; see, for example, [346, Corollary 3.3.16]. His proof involved quantifier elimination for the theory of

$$\mathbb{R}_{\text{alg}} = (\mathbb{R}, <, +, \times, 0, 1)$$

in the language $\mathcal{L}_{\text{rings}} \cup \{<\}$. Note that $<$ is definable in $(\mathbb{R}, +, \times, 0, 1)$, as noted above, but with quantifiers. So the inclusion of $<$ in the language adds no new definable sets but is needed for quantifier elimination. Decidability can be viewed as saying that the definable sets are not too complicated. A beautiful exposition of the result is in [151].

The theory of $\mathbb{C}_{\text{alg}} = (\mathbb{C}, +, \times, 0, 1)$ is also decidable, but this is easier. Via elimination theory it reduces essentially to algorithmic testing of multivariate ideal membership, and goes back to [275]; see also [483].

Tarski [499] asked whether the theory of the real field with the exponential function is decidable (in fact, he asked about the function 2^x but the functions are inter-definable, as noted in [186], so the questions are equivalent) in the corresponding language with an additional function symbol for $y = e^x$ (or one can equivalently add a relation symbol for the graph of exponentiation).

8.1 Question (Tarski [499, p. 45]) *Is the theory of the structure*

$$\mathbb{R}_{\text{exp}} = (\mathbb{R}, <, +, \times, 0, 1, e^x)$$

in its natural language decidable?

The discussion in van den Dries [186] observes several distinct aspects of Tarski's result, including quantifier elimination, structure of definable sets, coincidence of the theory with the theory of real closed fields, and decidability. He notes that quantifier elimination does not hold in $\mathbb{R}_{\exp}$ (by an old example; see e.g. [186]), that decidability would entail a lot of knotty arithmetic questions (e.g. whether $P(e, e^e, \ldots) = 0$ for each individual $P \in \mathbb{Z}[x_1, x_2, \ldots]$), but that these issues are separable from the question of whether the definable sets are nicely structured as unions of cells, which revolves around the key finiteness property we state shortly.

O-Minimal Structures

The finiteness property highlighted by van den Dries was developed in a general setting of expansions of an ordered structure by Pillay–Steinhorn [441, 442], who coined the term "o-minimal structure". The Cell Decomposition Theorem, which is the fundamental structural result describing definable sets in an o-minimal structure, is proved in Knight–Pillay–Steinhorn [312] and extended to a slightly more general setting in [443]. The main reference for o-minimality (and for our summary below) is [189].

The setting is an expansion of a dense linear order (assumed to be without endpoints, for simplicity), though the most interesting examples expand a field structure.

8.2 Definition   Let $\mathcal{R} = (R, <, \ldots)$ be an expansion of a dense linear order without endpoints. An *interval* in $\mathcal{R}$ is a set of the form

$$(a, b) = \{x \in R : a < x < b\},$$

where $a \in R \cup \{-\infty\}, b \in R \cup \{\infty\}$, and the inequalities $-\infty < x < \infty$ hold for all $x \in R$.

The following definitions set out what it means for a set $Z \subset \mathbb{R}^n$ or a function $f : A \to \mathbb{R}$ on some $A \subset \mathbb{R}^n$ to be *definable in an o-minimal structure*.

8.3 Definition ([189, §2, 3.2; 345, Definition 3.1.18])   An expansion $\mathcal{R} = (R, <, \ldots)$ of a dense linear order without endpoints is called *o-minimal* if every definable subset of R (with parameters) is a finite union of points and intervals.

8.4 Definition   Let $\mathcal{R} = (R, <, \ldots)$ be an o-minimal structure and $A \subset R^n$. A function $f : A \to R$ is said to be *definable in $\mathcal{R}$* if the graph of f as a subset of R^{n+1} is definable in $\mathcal{R}$ (whence the domain A is definable).

8.5 Remarks

1. Note that, in the context of o-minimality, *definability is nearly always taken to be with parameters*. We assume this henceforward.

2. As the endpoints of intervals are required by definition to be in R, any finite union of points and intervals is already definable (with parameters) in $(R, <)$. Thus, the o-minimality condition is that the collection of definable subsets of R is as small as it could be in the presence of $<$. Hence, the term o-minimal (short for order-minimal), the analogue of minimality in an ordered structure.

3. In R^2, there are in general plenty of sets not definable with only the order relation, but the point is that the restriction on the definable subsets of R exerts considerable control.

4. As an o-minimal structure is, formally, an expansion of an ordered structure, we include the order relation $<$ in the structures $\mathbb{R}_{alg}$ and $\mathbb{R}_{exp}$ even though it is definable in any expansion of the real field.

5. O-minimality steers far from definability of $\mathbb{Z}$ by excluding all discrete infinite sets. However, one can still have tame behaviour when certain such sets are admitted. See, for example, [190]. The boundary of tame behaviour in extensions of the real field is closely explored in [277].

Model-Theory Free Definition of O-Minimality

The collection of definable sets in a structure has certain properties flowing from the syntax of first-order languages. Since one can combine formulae using $\land, \lor, \neg, \rightarrow$, and $\leftrightarrow$, the collection of definable subsets of A^n is, for each n, a boolean algebra. And, due to the presence of quantifiers, the image of a definable set in A^n under some coordinate projection to A^m is definable. If the structure expands the real field $\mathbb{R}_{alg}$, then every semi-algebraic set is definable.

These properties enable a model-theory free definition of an o-minimal expansion of a field which is found in several places (e.g. [439]): an o-minimal structure over the real field $\mathbb{R}$ is a sequence $S = (S_n)$ of collections S_n of subsets of $\mathbb{R}^n$ for $n = 1, 2, \ldots$ satisfying the above-mentioned closure properties, containing all semi-algebraic sets, and having the minimality property for definable subsets of $\mathbb{R}$. Model-theoretically, this structure is then $\mathbb{R}_S = (\mathbb{R}, S)$.

This definition has the attractive features of brevity and of requiring little overhead, but it is also less intuitive. It is obvious, for example, that the set of points where a definable function is, for example, continuous, or differentiable, is definable in view of the standard definition $\forall \epsilon > 0 \exists \delta > 0$, and so on. But expressing this using projections and complements is an exercise.

Definable Families

8.6 Definition   A *definable family* of sets in $\mathcal{R}^n$ is a definable set $Z \subset \mathcal{R}^m \times \mathcal{R}^n$ viewed as the family $\{Z_y : y \in \mathcal{R}^m\}$ of its fibres $Z_y = \{x \in \mathcal{R}^n : (y, x) \in Z\}$ in $\mathcal{R}^n$.

8.7 Remarks

1. Note that we have put the parameter space to the left of the ambient space. This choice is important below as cells are defined with respect to a fixed ordering of the coordinates.
2. It is not required that the parameter space of the family be all of $\mathcal{R}^m$, but it is always a definable set (being the image of Z under projection to $\mathcal{R}^m$).

Cell Decomposition

The fundamental result about o-minimal structures is the Cell Decomposition Theorem [312].

8.8 Definition ([189, §3, Definition 2.3])   Let $\mathcal{R} = (R, <, \ldots)$ be an ordered structure, and $(i_1, \ldots, i_n) \in \{0, 1\}^n$. An $(i_1, \ldots, i_n)$-*cell* is a definable set in R^n defined inductively as follows:

1. A (0)-cell in R is a point; a (1) cell in $\mathbb{R}$ is an open interval (in the above sense).
2. An $(i_1, \ldots, i_n, 0)$-cell in R^{n+1} is the graph of a definable continuous function $f : C \to R$, where $C \subset R^n$ is an $(i_1, \ldots, i_n)$-cell; an $(i_1, \ldots, i_n, 1)$-cell in R^{n+1} is the region between the graphs of two definable continuous functions $f, g : C \to R$, where $C \subset R^n$ is an $(i_1, \ldots, i_n)$-cell and $f < g$ on C.
3. A cell is an $(i_1, \ldots, i_n)$-cell for some $(i_1, \ldots, i_n)$ that is unique and called the *type* of the cell.

Note that this definition fixes an explicit order of the coordinates $x_1, \ldots, x_n$; the notion is not stable under permutation of coordinates. A cell in $R^m \times R^n$ may be considered as a definable family with parameters in R^m, and then the projection onto R^m is a cell, and the fibres are also cells and of constant type.

8.9 Definition ([189, §3, Definition 2.10])   A *decomposition of* R^n is a partition of R^n into disjoint cells with a certain hereditary property, defined inductively as follows.

1. A decomposition of R is a finite set of disjoint cells that partition R.
2. A decomposition of R^{n+1} is a finite partition of R^{n+1} into cells such that the set of projections of these cells to R^n is a decomposition of R^n.
3. If $A \subset R^n$ is a set and D a decomposition of R^n, we say that D is *compatible* with A if every cell in D is either contained in or disjoint from A.

8.10 Theorem (Cell Decomposition Theorem; [189, §3, Theorem 2.11; 312])
Let $A_1, \ldots, A_k \subset R^n$ be definable in an o-minimal structure $(R, <, \ldots)$. Then, there is a decomposition of R^n compatible with $A_1, \ldots, A_k$.

8.11 Remarks

1. With $k = 1$, one sees that every definable set is a finite disjoint union of cells.
2. The cells forming the decomposition are definable sets.
3. Suppose $X \subset R^n$ is definable in an o-minimal structure $(R, <, \ldots)$. Then, it is definable in the reduct $(R, <, X)$, which is also o-minimal. Hence, X has a cell decomposition with cells definable in this structure. In general, one does need $<$.

The following Uniform Finiteness Theorem follows from Cell Decomposition; however, the two theorems are proved simultaneously by induction on n.

8.12 Definition Let $A \subset R^{n+1}$.

1. The set A is called *finite over R^n* if for each $x \in R^n$ the fibre $A_x = \{y \in R : (x, y) \in A\}$ is a finite set.
2. The set A is called *uniformly finite over R^n* if $\#A_x$ is uniformly bounded for all $x \in R^n$.

8.13 Theorem (Uniform Finiteness Theorem; [189, Lemma 2.13]) *Let $A \subset R^{n+1}$ be definable in an o-minimal structure $(R, <, \ldots)$. If A is finite over R^n, then it is uniformly finite over R^n.*

8.14 Remarks

1. A consequence of cell decomposition (specifically uniform finiteness) is that o-minimal and *strongly o-minimal* coincide: every model of the theory of an o-minimal structure is o-minimal as finiteness, being uniform, is coded into the theory.
2. The Cell Decomposition Theorem (or uniform finiteness) implies the zero estimate in the proof of Theorem 2.3. Indeed, for $Z \subset \mathbb{R}^2$ definable in any o-minimal structure expanding a field, and positive integer d, the intersections of Z with the family of real algebraic plane curves of degree d is a definable

family. For f real analytic, as in Theorem 2.3, its graph Z is definable in $\mathbb{R}_{an}$; see §8.21.

3. Note that a uniform zero estimate holds for Z varying in a definable family, and uniform finiteness for the number of isolated intersections still holds even if some fibres in the family have positive-dimensional intersections.

Properties

O-minimality has strong consequences for the definable sets. We just note some basic properties.

8.15 Proposition *Let $\mathcal{R} = (R, <, \ldots)$ be an o-minimal structure.*

1. *A definable function on a subset $Z \subset R$ is continuous at all but finitely many points.*
2. *Suppose $\mathcal{R}$ expands a field. A definable function on a subset $Z \subset R$ is differentiable at all but finitely many points.*

One can then refine the cell decomposition to ensure that the cell boundaries are C^r for any fixed r. Some o-minimal structures have analytic cell decomposition (e.g. $\mathbb{R}_{an,exp}$; see [192]), but others lack C^∞ cell decomposition ([333]).

Now the ordered field R is in general not Dedekind complete, and may be totally disconnected. It may be non-archimedean. However, definability imposes a tame topology.

8.16 Definition

1. ([189, Definition 3.5]) A set $X \subset R^n$ is called *definably connected* if it is definable but it is not the union of two disjoint non-empty definable open subsets of X.
2. ([189, p. 57]) A *definably connected component* of a non-empty definable set $X \subset R^n$ is a maximal definably connected subset of X.

By the following theorem, a non-empty definable set is the union of (and indeed is partitioned by) its finitely many definably connected components.

8.17 Theorem

1. ([189, Proposition 2.9]) *A cell is definably connected.*
2. ([189, Proposition 2.18]) *A non-empty definable set $X \subset R^n$ has only finitely many definably connected components. They are open and closed in X and form a finite partition of X.*

Dimension

Definable sets in an o-minimal structure have a well-behaved notion of dimension. We give the definition and a few key properties needed in what follows.

8.18 Definition ([189, §4, 1.1]) Let $Z \subset \mathcal{R}^n$ be definable in an o-minimal structure $(\mathcal{R}, <, \ldots)$. The *dimension* of Z is defined by

$$\dim Z = \max\{i_1 + \cdots + i_m : Z \text{ contains an } (i_1, \ldots, i_m)\text{-cell}\}.$$

The empty set is defined to have $\dim \emptyset = -\infty$.

If X is definable, then its interior $\mathrm{int}(X)$ and closure $\mathrm{cl}(X)$ are definable. The *boundary* of X is defined to be $\mathrm{cl}(X) \backslash \mathrm{int}(X)$ and the *frontier* of X is defined to be $\mathrm{fr}(X) = \mathrm{cl}(X) - X$. Then, $\dim \mathrm{fr}(X) < \dim X$, for any non-empty definable set X, and in particular $\dim \mathrm{cl}(X) = \dim X$ ([189, Ch. 4, Theorem 1.8]).

8.19 Proposition ([189, §4, Proposition 1.5]) *Let $Z \subset \mathcal{R}^m \times \mathcal{R}^n$ be a definable set. For $k \in \{-\infty, 0, 1, \ldots\}$, set*

$$Z(k) = \{y \in \mathcal{R}^m : \dim Z_y = k\}.$$

Then, $Z(k)$ is definable and the part $Z \cap Z(k) \times \mathcal{R}^n$ of Z over $Z(k)$ has dimension

$$\dim Z \cap Z(k) \times \mathcal{R}^n = \dim Z(k) + k.$$

Thus, in a definable family, the set of parameters for which the fibre has given dimension is definable. In view of this property, it is sometimes said that in an o-minimal structure "one can talk definably about dimension", though this is a bit imprecise.

Examples of O-Minimal Structures

An o-minimal structure need not expand a field structure (see examples in [189]), or may expand a field other than $\mathbb{R}$ (though the underlying field is always real closed). However, the most studied examples are expansions of the real field. Note that so far as the definable sets are concerned, it makes no difference whether the basic sets defining the structure are functions or their graphs.

8.20 Semi-algebraic Sets
The prototype example of an o-minimal structure is $\mathbb{R}_{\mathrm{alg}}$. Its o-minimality follows from the above-mentioned work of Tarski [499].

8.21 Globally Subanalytic Sets
van den Dries observed in [187] that the structure generated by *restricted analytic functions* is o-minimal as a consequence of a famous theorem of Gabrielov

[222]; see also [61]. An independent proof through quantifier elimination in an extended language (which thus recovers Gabrielov's Theorem) in the real and p-adic settings is given in [177]. The structure

$$\mathbb{R}_{an} = (\mathbb{R}, <, +, \times, 0, 1, \{Z\})$$

is generated by adding to the real field the graphs Z of all functions $f : [0, 1]^n \to \mathbb{R}$, where f is real analytic on some open neighbourhood of the unit cube (but note that the graph is then restricted to the unit cube; hence the term *restricted analytic*).

The definable sets in $\mathbb{R}_{an}$ are called *globally subanalytic* as they are subanalytic "at infinity"; that is, when considered as subsets of $\mathbb{P}^n(\mathbb{R})$ and to distinguish them from the subanalytic sets as studied, for example, in [61], in which the property is local (see also [187]).

The relationship between subanalytic sets as an *analytic-geometric category* extending the subanalytic category and o-minimal structures extending $\mathbb{R}_{an}$ is explored in [193], and detailed analytic properties are adduced.

8.22 Definition ([189, S2, Notes, p. 41])   An o-minimal structure is called *polynomially bounded* if every definable function on an interval (a, ∞) has growth bounded by some polynomial.

The structure $\mathbb{R}_{an}$ is polynomially bounded ([187]).

8.23 The Real Exponential
While $\mathbb{R}_{an}$ is quite rich, the key motivating question in the theory was whether the structure $\mathbb{R}_{exp} = (\mathbb{R}, <, +, \times, 0, 1, e^x)$ generated by the *full real exponential graph* is o-minimal. The unrestricted graph of exponentiation is not definable in $\mathbb{R}_{an}$ (e.g. it is not polynomially bounded).

The fact that the germs at infinity of *univariate* functions built from exp and log and algebraic functions form an ordered field (and so such functions have only finitely many zeros) goes back to Hardy [265]. Finiteness of connected components for sets defined by exponential algebraic equations follows from Khovanskii's theory of Pfaffian systems of differential equations in connection with the theory of "Fewnomials" ([303]). But o-minimality entails more due to the quantifiers (which cannot be eliminated; see e.g. [186]). The model-completeness of the (theory of the) structure $\mathbb{R}_{exp}$, implying its o-minimality, is established by Wilkie in [527]; see also [130] (and [407]).

8.24 Theorem (Wilkie [527])   *The theory of $\mathbb{R}_{exp}$ is model-complete; hence $\mathbb{R}_{exp}$ is o-minimal.*

The model-completeness of $\mathbb{R}_{\exp}$ entails the following explicit form of definable sets (see e.g. [293]).

8.25 Theorem *A set $Z \subset \mathbb{R}^n$ is definable in $\mathbb{R}_{\exp}$ if and only if there exists a positive integer m and a real polynomial in $2n + 2m$ variables such that Z is the projection onto the first n variables x in $(x, y) = (x_1, \ldots, x_n, y_1, \ldots, y_m)$ of*

$$\{(x, y) \in \mathbb{R}^{n+m} : Q(x_1, \ldots, x_n, y_1, \ldots, y_m, e^{x_1}, \ldots, e^{x_n}, e^{y_1}, \ldots, e^{y_m}) = 0\}.$$

O-minimality follows from model-completeness by the aforementioned results of Khovanskii.

The Structure $\mathbb{R}_{an,\exp}$

In [192], van den Dries and Miller show that the structure whose definable sets are generated by those of both $\mathbb{R}_{an}$ and $\mathbb{R}_{\exp}$ is (model-complete and) o-minimal.

8.26 Theorem ([192]) *The structure*

$$\mathbb{R}_{an,\exp} = (\mathbb{R}, <, +, \times, 0, 1, \exp, \{Z\}),$$

where $\{Z\}$ is the set of all graphs of restricted analytic functions, is model-complete and o-minimal.

O-minimality is proved (by a different approach, through quantifier elimination in an expanded language) in [191].

All the sets required in the diophantine applications described in this book are definable in $\mathbb{R}_{an,\exp}$ (and of course in some much smaller reduct of it); see Ingredient 6.10. For example, $e \colon \mathbb{C} \to \mathbb{G}_m$ restricted to the domain $F = [0, 1] \times i\mathbb{R}$ is definable there using the full real exponential and the restrictions of sine and cosine. Hence, the set Z in Ingredient 3.10 is definable (**exercise**). As another example, via its q-expansion, $j(z)$ is an analytic function of $q = e(z)$ that is definable on the fundamental domain (the q-expansion is over-convergent as the fundamental domain is bounded away from the real axis). Hence, the set Z in Ingredient 5.3 is definable in $\mathbb{R}_{an,\exp}$.

Larger o-minimal structures exist (see below). However, it can rather be useful to show that a given set is definable in a smaller o-minimal structure, which might have better model-theoretic properties or for which effective or explicit bounds might be available. For example, many of the improved results for counting (and applications) have been established in the setting of Pfaffian functions and the associated o-minimal structure $\mathbb{R}_{pfaff}$ (see below), though note that $\mathbb{R}_{pfaff}$ is not a reduct (smaller structure) of $\mathbb{R}_{an,\exp}$.

8.27 Definition   The reduct of $\mathbb{R}_{an,exp}$ containing $\mathbb{R}_{an}$ and all power functions $x \mapsto x^s, s \in \mathbb{R}$ for $x \in (0, \infty)$ is denoted by $\mathbb{R}_{an}^{pow}$.

8.28 Pfaffian Functions

The notion of a Pfaffian function is due to Khovanskii [303], who showed that the zero sets defined by systems of such functions have only finitely many connected components (see also [344]). Explicit general bounds have been given, and for more general types of sets defined using Pfaffian functions, by Khovanskii, Gabrielov–Vorobjov, and others, see, for example, [80, 223]. For general definable sets, one has primitive recursive bounds [46].

8.29 Definition   Let $U \subset \mathbb{R}^n$ be an open domain.

1. A sequence of real analytic functions $f_1, \ldots, f_r : U \to \mathbb{R}$ is called a *Pfaffian chain of order r and degree α* if, for each i, j with $1 \le i \le r$ and $1 \le j \le n$, we have $\partial f_i / \partial x_j = P_{ij}(x_1, \ldots, x_n, f_1, \ldots, f_i)$ for some $P \in \mathbb{R}[X_1, \ldots, X_n, Y_1, \ldots, Y_i]$ with $\deg P_{i,j} \le \alpha$.
2. A function $f : U \to \mathbb{R}$ is called *Pfaffian function of order r and degree (α, β)* if there is a Pfaffian chain $f_1, \ldots, f_r$ of order r and degree α and a polynomial $Q \in \mathbb{R}[X_1, \ldots, X_n, Y_1, \ldots, Y_r]$ of degree $\deg Q \le \beta$ such that $f = Q(x_1, \ldots, x_n, f_1, \ldots, f_r)$.
3. By a *total Pfaffian function* we mean a Pfaffian function defined using a chain with $U = \mathbb{R}^n$.

One sometimes sees this definition with the weaker hypothesis that the f_i are C^1. However, it follows from the condition that all the f_i are real analytic (see [192, 8.2]).

8.30 Definition   Let F be the set of all total Pfaffian functions. Define

$$\mathbb{R}_{pfaff} = (\mathbb{R}_{alg}, F).$$

8.31 Theorem (Wilkie [529])   *The structure $\mathbb{R}_{pfaff}$ is o-minimal.*

The proof is quite different to that of $\mathbb{R}_{exp}$ in [527], in which o-minimality follows from model-completeness in view of the results in [303]. Model-completeness of $\mathbb{R}_{pfaff}$ is not known.

8.32 Definition   The reduct of $\mathbb{R}_{an}$ in which one expands the real field by the graphs of the restriction of Pfaffian functions to a compact box $B \subset U$ is denoted as $\mathbb{R}_{respfaff}$.

The structure $\mathbb{R}_{respfaff}$ enjoys good properties coming from both $\mathbb{R}_{an}$ and $\mathbb{R}_{pfaff}$. An even smaller reduct of *restricted elementary functions* has even better properties as the defining functions have the property that their complexifications are Pfaffian when considered in real variables, enabling a combination of complex analytic and Pfaffian techniques to be deployed in [76].

8.33 Definition ([188]) The structure of *restricted elementary functions* is given by

$$\mathbb{R}_{RE} = \big(\mathbb{R}, <, +, \times, 0, 1, \exp |_{[0,1]}, \sin |_{[0,\pi]}\big).$$

8.34 Further Examples

Various authors have constructed o-minimal structures larger than $\mathbb{R}_{an,exp}$. The constructions in [462] exhibit structures that do not admit analytic cell decomposition, and show the existence of incompatible pairs of o-minimal structures (two o-minimal structures such that the structure generated by the union of their collections of definable is not o-minimal). In particular, there is no largest o-minimal structure.

Any o-minimal structure $\mathcal{R}$ over the real field has a *Pfaffian closure* $\mathcal{P}(\mathcal{R})$ that is o-minimal ([492]). Note that this is a more general construction than that of $\mathbb{R}_{pfaff}$ above (which is generated by total Pfaffian functions) and $\mathcal{P}(\mathbb{R}_{alg})$ might be larger than $\mathbb{R}_{pfaff}$; see [297]. Any o-minimal structure over $\mathbb{R}$ that is not polynomially bounded defines the full exponential function ([370]). (The same holds, appropriately formulated, over any o-minimal expansion of a field; see [371].)

All the above structures are *exponentially bounded* (all definable functions on an interval (a, ∞) are eventually bounded by an iterated exponential; see [192, 336]). The existence of a *transexponential* o-minimal structure is unknown, though there do exist Hardy fields containing transexponential functions [95].

Every o-minimal structure over $\mathbb{R}$ has non-standard models. For exotic examples of non-archimedean o-minimal structures that do not arise in this way, and exhibit unusual features, see [284]. See also the discussion of non-standard tori and elliptic curves in [411].

Finiteness Ingredient in 3.11 and 5.13

We can now show how o-minimality, in conjunction with Ax–Lindemann for the modular function, implies the Finiteness Ingredient, 5.13, in the proof of MAO. We prove it in a stronger form, for families. The proof of 3.11 for families is similar.

8.35 Proposition *Let $V \subset Y(1)^n \times P$ be a family of algebraic subvarieties parameterized by $t \in P = P(\mathbb{C})$ for some quasi-projective variety P. There exists a finite number of families of weakly special subvarieties such that, for* *every $t \in P$, every maximal weakly special subvariety $W \subset V_t$ is a fibre in one of these families.*

Proof Let $t \in P$. The maximal algebraic subvarieties contained in $j^{-1}(V_t)$ are weakly special subvarieties, and hence are Möbius subvarieties. Thus, they are also maximal among Möbius subvarieties contained in $j^{-1}(V_t)$, and vice versa: the maximal Möbius subvarieties contained in $j^{-1}(V_t)$ are weakly special. Further, every Möbius subvariety is $SL_2(\mathbb{Z})^n$-equivalent to one that passes through F^n.

Möbius subvarieties come in finitely many families, depending on the choice of a strict partition p. Hence, the entire collection of Möbius subvarieties of $\mathbb{H}^n$ is a definable family (in $\mathbb{R}_{alg}$). We form the definable set of $SL_2(\mathbb{R})$ parameters for Möbius varieties that have a translation (i.e. a choice of the fixed coordinates) for which the corresponding Möbius variety intersects F^n in a set of full dimension (the number k of parts p_i, $i \geq 1$, in the strict partition), where, for some $t \in P$, it is contained in the definable set $j^{-1}(V_t) \cap F^n$ and is maximal over all Möbius subvarieties with these properties, for that t. This set then consists entirely of points representing the slopes (see Definition 4.14.6) of the maximal weakly special subvarieties of the V_t, and every maximal weakly special subvariety of every V_t has its slope represented in this set. These representatives lie in the image of $GL_2^+(\mathbb{Q})$ in $SL_2(\mathbb{R})$, which cannot contain an interval of positive dimension. Therefore, in the cell decomposition of this definable set there are no positive-dimensional cells. Hence, this set is finite.   □

8.36 Proof of 3.11
As above, but using translates of real linear subspaces in place of Möbius varieties.   □

8.37 Remark
The image of $GL_2^+(\mathbb{Q})$ in $SL_2(\mathbb{R})$ is countable, whereas positive-dimensional cells in an o-minimal structure over $\mathbb{R}$ are uncountable. So here it suffices to note the countability; but in general the ambient set of an o-minimal structure may itself be countable, so that the dictum "a countable definable set is finite" is not true in general.

p-Adic Analogues

Various analogues of o-minimality have been studied in the p-adic numbers, or more generally in valued fields. Real and p-adic subanalytic sets are studied in parallel in [177]. The notion of P-minimality, for p-adically closed fields, is introduced in [267]. See [144] for the notion of Hensel minimality (in Henselian valued fields) and discussion, and [117, 145] for diophantine applications.

What about Tarski's Question?

The above sets out the basic notions and structures we require. To conclude the historical story, we mention that Macintyre–Wilkie [340] (see also [528]) answer Tarski's question in the affirmative, conditionally on (a property implied by) Schanuel's conjecture (Conjecture 13.1).

8.38 Theorem ([340])   *Assume Schanuel's conjecture. Then, $\mathbb{R}_{exp}$ is decidable.*

9

Parameterization and Point-Counting

r-Parameterizations

The key idea in proving the Counting Theorem is to show that definable sets in o-minimal expansions of a field can be parameterized in a nice way, uniformly in families.

9.1 Definition Let $Z \subset (0,1)^n$, with dim $Z = k$. By an *r-parameterization* of Z we mean a finite collection $\Phi = \{\phi\}$ of maps

$$\phi \colon (0,1)^k \to (0,1)^n, \quad \phi = (\phi_1, \ldots, \phi_n), \quad \phi_i \in C^r\left((0,1)^k\right)$$

for $i = 1, \ldots, n$ such that

$$Z = \bigcup_{\phi \in \Phi} \phi\left((0,1)^k\right)$$

(i.e. they parameterize Z), with all the coordinate functions ϕ_i possessing derivatives up to order r and *all partial derivatives up to order r bounded by* 1 *in absolute value.*

Yomdin [537] and Gromov [248] introduced r-parameterizations (which are also called C^r-parameterizations, and are sometimes defined using closed boxes) of semi-algebraic sets, and what is known as the Yomdin–Gromov Algebraic Lemma, as a key device in the proof by Yomdin [536] of Shub's entropy conjecture for C^∞ maps. The result is that given a closed algebraic set $A \subset [0,1]^n$, and $r \geq 1$, A has an r-parameterization in which the number of parameterizing functions depends only on r and the dimension and degree of (the polynomials defining) A. Yomdin's original result allowed parameterizations that left out small pieces, and the result parameterizing the full set is in the form given by Gromov.

The proof of the Counting Theorem depends on the generalization of the result to definable families in an arbitrary o-minimal structure. The proof in

82

[248] is quite abbreviated. A detailed proof ([111]) in the semi-algebraic case appeared around the same time time as the generalization in [439].

9.2 Definition Let $Z \subset P \times \mathbb{R}^n$ be a definable family of sets in an o-minimal expansion of $\mathbb{R}$, and $r \geq 1$ an integer. A *definable r-parameterization* of Z is a finite set Φ of definable families

$$\phi : P \times (0,1)^k \to (0,1)^n$$

of maps $(0,1)^k \to (0,1)^n$ such that, for each $y \in P$, the finite set of fibres $\{\phi_y\} : \phi \in \Phi$ is an r-parameterization of Z_y.

9.3 Theorem (The r-Parameterization Theorem; [439]) *Let Z be a definable family of sets in $(0,1)^n$, and $r \in \mathbb{N}$. Then, there exists a definable r-parameterization of Z.*

9.4 Remark The same statement holds in any o-minimal expansion of a field.

The key ideas in the Yomdin–Gromov construction are elementary and can be adapted to the o-minimal setting. For functions of one variable the main devices are

(i) that if derivatives up to some order are bounded and the next derivative is monotonic, then the mean value theorem imposes a bound on how it degenerates;

(ii) this degeneration can be made bounded by the simple reparameterization $x \mapsto x^2$ near the origin; in conjunction with

(iii) one can divide into subintervals to achieve (differentiability up to a given order and) monotonicity of derivatives, and the number of subdivisions is uniformly bounded in families, by o-minimality. In several variables one

(iv) reparameterizes the function giving the maximum of a partial derivative over a range of other variables. This requires some care when the maximum loses smoothness. These ingredients are combined in a careful induction.

See the expository account [531], a more elementary proof in [313], and a proof along the original lines of [248] in [78].

A Determinant Estimate

We now give an analogue of Corollary 2.5 for functions of several variables. It shows that an alternant evaluated in a small disc is small. For functions of several variables there is no Vandermonde to divide by, so we do not have an analogue of the mean value theorem for alternants (Lemma 2.4). Instead, the idea is to expand the entries in Taylor series with the remainder in order to register cancellation of low-order terms. Then the remaining terms have high orders and are small. But one also does not want to have too many terms, and this guides the order of the expansion.

We follow the exposition in [421] and start with setting up some notation and stating a multivariate version of Taylor's Theorem with remainder, following [316]. We do this for real functions, but it all goes through in an arbitrary o-minimal expansion of a real closed field; see, for example, [189, Exercise 2.12.2].

9.5 Notation For $k \in \mathbb{N}$ and $\mu = (\mu_1, \ldots, \mu_k) \in \mathbb{N}^k$, we set $|\mu| = \mu_1 + \cdots + \mu_k$ and $\mu! = \mu_1! \cdots \mu_k!$. For $x = (x_1, \ldots, x_k)$, we write x^μ for $x_1^{\mu_1} \cdots x_k^{\mu_k}$ and

$$\frac{\partial^\mu}{\partial x^\mu} = \frac{\partial^{\mu_1}}{\partial x_1^{\mu_1}} \cdots \frac{\partial^{\mu_k}}{\partial x_k^{\mu_k}}.$$

With this notation, suppose that $\phi \colon \mathbb{R}^k \to \mathbb{R}$ is defined and has $b + 1$ continuous derivatives at each point of a region containing the line segment joining $y, z \in \mathbb{R}^k$. By Taylor's formula (see e.g. [316, Theorem 2.2.5]), there is a point ξ in this line segment such that

$$\phi(z) = \sum_{\{\mu \in \mathbb{N}^k : |\mu| \le b\}} \frac{1}{\mu!} \frac{\partial^\mu}{\partial x^\mu} \phi(y)(z - y)^\mu + \sum_{\{\mu \in \mathbb{N}^k : |\mu| = b+1\}} \frac{1}{\mu!} \frac{\partial^\mu}{\partial x^\mu} \phi(\xi)(z - y)^\mu.$$

Now, for $k, d \in \mathbb{N}$ set

$$L_k(d) = \#\{\mu \in \mathbb{N}^k : |\mu| = d\}, \quad D_k(d) = \#\{\mu \in \mathbb{N}^k : |\mu| \le d\}.$$

Then we have

$$L_{k+1}(d) = D_k(d) = \binom{k+d}{k}, \quad D_k(d) = \sum_{\delta=0}^{d} L_k(\delta).$$

Thus, for fixed k, $L_k(d) \in \mathbb{Q}[d]$ with degree $k - 1$ and leading coefficient $1/(k-1)!$, while $D_k(d) \in \mathbb{Q}[d]$ with degree k and leading coefficient $1/k!$.

We now consider a $D \times D$ alternant formed using the functions $\psi_1, \ldots, \psi_D \colon \mathbb{R}^k \to \mathbb{R}$. In the intended application, the functions ψ_i are the monomial functions in n variables on the ϕ_i up to some degree d, so that $D = D_n(d)$. Now we choose the order of Taylor expansion b to fit this application.

9.6 Definition Given $k, n, d \in \mathbb{N}$, there is a unique $b = b(k, n, d)$ such that $D_k(b) \le D_n(d) < D_k(b + 1)$. Set

$$B(k, n, d) = \sum_{\beta=0}^{b} L_k(\beta) \beta + \left(D_n(d) - \sum_{\beta=0}^{b} L_k(\beta) \right)(b + 1).$$

9.7 Lemma ([421, Lemma 3.1]) *Let $k, n, d \in \mathbb{N}$. Put $D = D_n(d)$, $b = b(k, n, d)$, and $B = B(k, n, d)$. Let $J \subset \mathbb{R}^k$ be a convex domain and let $\psi_1, \ldots, \psi_D : J \to \mathbb{R}$ be functions possessing at least $b + 1$ derivatives that are continuous and bounded in absolute value on J. There is a constant $c(J, \phi_1, \ldots, \phi_D)$ with the following property. Let $U \subset \mathbb{R}^k$ be a disc of radius $r \leq 1$ and $z^{(1)}, \ldots, z^{(D)} \in U \cap J$. Then,*

$$\left| \det \left(\psi_i(z^{(j)}) \right) \right| \leq c(J, \psi_1, \ldots, \psi_D) \, r^B.$$

Proof The intersection $J \cap U$ is a convex set and we can find a point $z^{(0)} \in J \cap U$ such that every other point there is at a distance at most $2r$. Expand each entry in a Taylor series with remainder of order $b + 1$ about $z^{(0)}$.

In expanding the determinant, consider the terms corresponding to a particular specification of the number of terms of each order of derivative. Consider a minor of $\det \left(\psi_i(z^{(j)}) \right)$ of size $h \times h$ involving a certain choice of terms in the Taylor expansion of degree β. That is, choose h points $\zeta^{(i)}$ from among the $z^{(j)}$, h functions g_i from among the ψ_j, and h monomials $v_i \in \mathbb{N}^k$ of degree β and consider

$$\det \left(\frac{1}{v_j!} \frac{\partial^{v_j}}{\partial x^{v_j}} g_j(z^{(0)}) \left(\zeta^{(i)} - z^{(0)} \right) \right).$$

If $h > L_k(\beta)$, then the columns are linearly dependent and the determinant of the minor vanishes.

Thus, if, for a particular specification of orders, there are more than $L_k(\beta)$ terms of order β for any $\beta = 0, \ldots, b$, then the totality of terms corresponding to this choice vanishes. Therefore, every surviving term has total order at least $B(k, n, d)$, and so

$$\left| \det \left(\psi_i(z^{(j)}) \right) \right| \leq c \, r^B,$$

where c depends on the maximum sizes of the derivatives of the ψ_i up to order $b + 1$ on $J \cap U$. □

Exploring with Hypersurfaces

Suppose we have a set $Z \subset (0, 1)^n$ of dimension k parameterized as the image of a function

$$\phi : (0, 1)^k \to (0, 1)^n$$

with sufficiently smooth coordinate functions $(\phi_1, \ldots, \phi_n)$.

We apply Lemma 9.7 using the $D = D_n(d)$ monomial functions ψ_i of degree d on $\phi_1, \ldots, \phi_n$, and taking D points $z^{(i)} \in (0, 1)^k \cap U$ the pre-images of points in $Z(\mathbb{Q}, T)$ for some disk U of radius $r \leq 1$. Set

$$\Delta = \det \left(\psi_j(z^{(i)}) \right).$$

Then Δ is a rational number with denominator at most $T^{dD_n(d)}$ and we have

$$dD_n(d) = \frac{1}{n!} d^{n+1} (1 + o(1)),$$

when $d \to \infty$ with k, n fixed. Now, in the same limit,

$$b = b(k, n, d) = \left(\frac{k! d^n}{n!}\right)(1 + o(1)).$$

Thus

$$|\Delta| \le cr^B,$$

where c depends on the sizes of the derivatives of the ϕ_i up to order $b + 1$ and

$$B = B(k, n, d) = \frac{1}{(k+1)(k-1)!} \left(\frac{k!}{n!}\right)^{(k+1)/k} d^{n(k+1)/k} (1 + o(1)).$$

Let

$$\epsilon = \epsilon(k, n, d) = \frac{dD_n(d)}{B(k, n, d)}.$$

Then the points in $Z(\mathbb{Q}, T)$ in the image of U all lie on one hypersurface of degree d, provided that

$$r \ll T^{-\epsilon},$$

and $(0, 1)^n$ may be covered by $\ll T^{k\epsilon}$ such balls.

The crucial point is that, k, n being fixed, $\epsilon(k, n, d) \to 0$ as $d \to \infty$, essentially because there are more monomials in n variables than in k variables when $k < n$. This is summarized in the following statement.

9.8 Proposition ([421, Proposition 4.2])　*Let Z be as above and $\epsilon > 0$. Then, for suitable $d = d(k, n, \epsilon)$, we have that $Z(k, T)$ is contained in at most*

$$c(\phi, \epsilon) T^\epsilon$$

hypersurfaces of degree d.

Sketch Proof of the Counting Theorem

Here we sketch the proof of the basic Counting Theorem 2.10 before giving a detailed proof of a refined version, which requires substantial preparation.

9.9 Sketch Proof　We start with a definable set (or family of sets) $Z \subset \mathbb{R}^n$.

Reflect into $(0, 1)^n$. By reflection and inversion $z \mapsto \pm z^{\pm 1}$ of coordinates (which preserve definability and height), and viewing intersections with $x_i = 1$ as subsets of $\mathbb{R}^{n-1}$, it suffices to consider $Z \subset (0, 1)^n$. By an r-parameterization we reduce to the case in which Z is the image of a function $\phi : (0, 1)^k \to (0, 1)^n$, where ϕ has sufficiently many derivatives, all bounded by 1 in absolute value.

Explore with hypersurfaces. By choosing a suitable degree d, and then a suitable r, r-parameterizing the family, and applying Proposition 9.8, we see that $Z(\mathbb{Q}, T)$ is contained in $\ll T^{\epsilon/2}$ intersections $Z \cap W$, where W is a hypersurface of degree d.

Iteration. Now we have a definable family of intersections $Z \cap W$ of our original set (or family) Z with all hypersurfaces W of degree d, parameterized by the coefficients of the defining equation. To iterate the argument we r'-parameterize these sets for some suitable r'. Thus, it is essential to be able to work in families even starting out with a single set Z. We can iterate the argument provided that the dimension of the fibres $Z \cap W$ decrease.

The algebraic part. How does the algebraic part manifest itself in this process? The above sketch oversimplifies a bit. One wants to say that where the dimension of $Z \cap W$ is not lower than that of W, one has found something in the algebraic part. This is true if $\dim W = \dim Z$ (see the proof of Theorem 9.14). Thus, rather than intersect with hypersurfaces, one intersects with algebraic varieties of dimension k by taking hypersurfaces in each $k + 1$ subset of coordinates. Then, when one intersects, pieces that remain of dimension k are indeed in the algebraic part. Thus, iterating, one either reduces dimension or produces sets in the algebraic part, ending up finally with at most $\ll T^{\epsilon}$ points.

Refinements and Proofs

We observe some refined versions of the Counting Theorem. These are essentially more careful recordings of what is achieved in the proof.

First, the set Z^{alg} may not be definable. For example, this is the case for

$$Z = \{(x, y, z) \in (2, \infty) : z = x^y\}.$$

However, we show that for each ϵ it is only necessary to exclude a *definable* subset of Z^{alg}, which depends on ϵ, to achieve the T^{ϵ} estimate, and moreover this may be done uniformly in families. This gives the following refinement of Theorem 2.13 (the version with $k = 1$ is [439, Theorem 1.10]).

9.10 Theorem ([424]) *Let* $Z \subset \mathbb{R}^m \times \mathbb{R}^n$ *be a definable family of sets, $k \geq 1$, and $\epsilon > 0$. There exists a definable family of sets $U = U(Z, k, \epsilon) \subset \mathbb{R}^m \times \mathbb{R}^n$ and a constant $c(Z, k, \epsilon)$ such that for each parameter y,*

$$U_y \subset (Z_y)^{\mathrm{alg}}$$

and

$$N(k, Z_y - U_y, T) \le c(k, Z, \epsilon)T^{\epsilon}.$$

The fibres of W_y are those pieces of the algebraic part of the fibre of Z that are found by the intersection with algebraic subvarieties up to some degree depending on ϵ. More precisely, the fibre W_y is the union over $\ell = 1, \ldots, L$, where L depends on ϵ, of those parts of the intersection of Z_y with an algebraic subvariety of dimension ℓ that have dimension ℓ. A by-product of the proof is an important refinement of the statement. Rather than excluding all of Z^{alg} and counting point on the complementary transcendental part of the set Z, the proof actually produces $\ll T^{\epsilon}$ sets that are either points or higher-dimensional definable sets that are "locally semi-algebraic".

This idea is made precise in [424, 3.2] with the notion of *blocks*. We do not repeat the definition, as we give a similar (but more flexible) notion here. Roughly, they are locally semi-algebraic sets whose boundary is definable but maybe not semi-algebraic, for example the region of the plane between the x-axis and the graph of $y = e^x$. The original notion was a bit cumbersome and yet did not carry enough information for some applications. A more refined notion of *quasi-algebraic* cell is introduced in [260], and here we introduce another variant called *underalgebraic cell*. An underalgebraic cell is a block in the sense of [424, 3.2] (and even a *basic block*), and thus Theorem 9.14 strengthens (in a marginal way) the counting result there, but need not be a quasi-algebraic cell, due to the non-singularity requirement on the latter.

9.11 Definition

1. An *underalgebraic cell* is a cell that is contained in an $\mathbb{R}^{\mathrm{alg}}$-definable cell of the same dimension.
2. A *family of underalgebraic cells* is a family $W \subset \mathbb{R}^m \times \mathbb{R}^n$ of cells and a semi-algebraic family $Y \subset \mathbb{R}^m \times \mathbb{R}^n$ such that every fibre of W is contained in a fibre of Y of the same dimension.

The use of underalgebraic cells (or blocks) is crucial in some applications (such as in Proposition 21.10). Below it is used to count algebraic points of bounded degree, but for that purpose they can be eliminated by using auxiliary polynomials instead of determinants (this is much more efficient as it does not increase the dimension of the set whose points are to be counted). Points of bounded degree may be parameterized by the polynomials they satisfy over $\mathbb{Q}$, hence reducing to rational points on an associated definable set. But the set of polynomials with real coefficients that have a given number as a root is positive-dimensional semi-algebraic. Therefore, the associated set, while definable, is equal to its algebraic part, and versions of the Counting Theorem that throw away the algebraic part before counting become trivial.

9.12 Definition Let k be a positive integer. For a real number α we define the *k-polynomial height* $H_k^{\text{poly}}(\alpha)$ as follows: If $[\mathbb{Q}(\alpha) : \mathbb{Q}] > k$, set $H_k^{\text{poly}}(\alpha) = \infty$. Otherwise

$$H_k^{\text{poly}}(\alpha) = \min \left\{ H(a) : a = (a_0, \dots, a_k) \in \mathbb{Q}^{k+1} \setminus \{(0, \dots, 0)\}, \sum_{j=0}^{k} a_j \alpha^j = 0 \right\}.$$

For $\alpha \in \mathbb{R}^n$, set $H_k^{\text{poly}}(\alpha) = \max \left\{ H_k^{\text{poly}}(\alpha_i), i = 1, \dots, n \right\}$.

We note the following relation between polynomial height and absolute multiplicative height, as observed in [424, between 5.1 and 5.2].

9.13 Proposition *Let* $k \geq 1$, $\alpha \in \mathbb{R}$ *with* $H_k^{\text{poly}}(\alpha) < \infty$. *Then,*

$$H_k^{\text{poly}}(\alpha) \leq 2^k H(\alpha)^k.$$

9.14 Theorem *Let* $Z \subset \mathbb{R}^m \times \mathbb{R}^n$ *be a definable family,* $k \geq 1$, *and* $\epsilon > 0$.

There exists a definable family $W = W(Z, k, \epsilon) \subset \mathbb{R}^M \times \mathbb{R}^m \times \mathbb{R}^n$ *of underalgebraic cells with the property that every fibre* $W_{(u,y)}$ *of* W *is a subset of* Z_y *(and so the positive-dimensional fibres are contained in* Z_y^{alg}*) and such that for every parameter* y *and every* $T \geq 1$, $Z_y(k, T)$ *is contained in the union of at most* $c(Z, k, \epsilon)T^\epsilon$ *fibres of* W_y.

Observe that, in this version, the Counting Theorem looks like a crude analogue of Multiplicative Manin–Mumford, in which all the rational points are contained in $\ll_{Z, \epsilon} T^\epsilon$ underalgebraic cells.

Proof of Theorem 9.14 Following [424], given Z and k, we form a new definable family X whose rational points are the coefficients of the minimal polynomials of algebraic points of degree at most k on fibres of Z. In view of Proposition 9.13 it suffices to prove the theorem using the k-polynomial height rather than H.

In order to get the conclusion in the required form we begin with some preparation involving only semi-algebraic sets and their cell decompositions. Then we introduce the definable family and make suitable further cell decompositions. Finally we apply Proposition 9.8.

Thus, we let $N = n(k + 1)$ and define the semi-algebraic set $B^* \subset \mathbb{R}^N \times \mathbb{R}^n$ as

$$B^* = \left\{ \left(t_j^{(i)}, z_i \right) \in \mathbb{R}^N \times \mathbb{R}^n : \quad i = 1, \dots, n, \quad j = 0, \dots, k : \right.$$

$$\left. \left(\bigwedge_{i=1}^n \neg \bigwedge_{j=0}^k \left(t_j^{(i)} = 0 \right) \right) \wedge \bigwedge_{i=1}^n \left(\sum_{j=0}^k t_j^{(i)} z_i^k = 0 \right) \right\}.$$

Over each point of $\mathbb{R}^N$ there are at most k points of $\mathbb{R}^n$, and therefore we may decompose B^* into a finite number of pairwise disjoint semi-algebraic sets B such that each is a graph over the image A of its projection to $\mathbb{R}^N$, the image of projection to $\mathbb{R}^n$ we denote C.

Fix one of the sets B and its projections A, C. We further decompose A into finitely many subsets according to whether each coordinate is equal to $-1, 0, 1$ or lies in one of the intervals $(-\infty, -1), (-1, 0), (0, 1), (1, \infty)$. Hence, by a finite decomposition and maps of the form $t \mapsto \pm t^{\pm 1}$, we may assume that

$$B \subset \mathbb{R}^n \times (0, 1)^N$$

for some (possibly smaller) N, with C the image under projection to $\mathbb{R}^n$, and A its image in $(0, 1)^N$.

We eventually take a definable family $X \subset \mathbb{R}^m \times (0, 1)^N$ for each such A and explore it with algebraic subvarieties of dimension k and suitably chosen degree $d = d(n, k, \epsilon/2)$. The choice of degree does not depend on X but only on (n, k). For now, however, we stay within the semi-algebraic setting.

We consider the family

$$V \subset \mathbb{R}^M \times (0, 1)^N$$

of semi-algebraic varieties of dimension k and degree d formed by imposing an equation on every set of $k + 1$ coordinates. We formally consider it as family

$$V \subset \mathbb{R}^M \times (0, 1)^N$$

with no dependence on the $\mathbb{R}^m$ parameters.

This gives a family $V \cap A$ of intersections of the fibres of V with A, and their in images C form a family $W \subset \mathbb{R}^M \times \mathbb{R}^m \times C$. We take a cell decomposition $\{C_i^W\}$ of this family.

Now associated with the definable family Z is a family

$$Y^* = \{(y, t_j^{(i)}, z_i) \in \mathbb{R}^m \times B^* : (y, z) \in Z\}.$$

After intersecting with the various octants and transforming by $t \mapsto \pm t^{\pm 1}$, we get a finite number of definable families

$$Y \subset \mathbb{R}^m \times (0,1)^N$$

such that union of their images in $\mathbb{R}^n$ is (fibrewise) Z, and each Y has its family of projections to $(0,1)^N$

$$X \subset \mathbb{R}^m \times (0,1)^N.$$

Let Z' be the fibre-wise image of the family Y in $\mathbb{R}^n$. It is a subset of the original Z, and of C. It is a definable family, and finitely many of such families have fibre-wise union the original family Z. So we can rename Z' as Z and it suffices to prove the theorem for this Z.

Now we also have the family of intersections of (fibres of) X with (fibres of) V, which we write as

$$X \cap V \subset \mathbb{R}^M \times \mathbb{R}^n \times (0,1)^N,$$

and the family of images as

$$Z \cap W \subset \mathbb{R}^M \times \mathbb{R}^m \times C,$$

and we can cell-decompose this family as $\{C_i^{X \cap W}\}$ compatibly with $\{C_i^W\}$.

The set of parameters $(u,y) \in \mathbb{R}^M \times \mathbb{R}^m$ such that some cell of the fibre of $Z \cap W$ is an underalgebraic fibre of a cell in the fibre of W is definable, and those of positive dimension are then contained in the algebraic parts of the fibre of Z.

Degree k points on Z of k-polynomial height T come from rational points on X of height at most T.

Now, given $y \in \mathbb{R}^m$ and $T \geq 1$, we subdivide $(0,1)^N$ into at most $c(n,k,\epsilon/2)T^\epsilon$ boxes in each of which the points of $X_y(k,T)$ are contained in one fibre of W. Each gives finitely many cells in $Z_y \cap W$. Some of these are underalgebraic. The others are fibres in a definable family of fibre dimension at most $k-1$.

The proof of Theorem 9.14 is completed by induction on the fibre dimension of X, in the course of which one finds $Z(k,T)$ contained in at most $c(Z,k,\epsilon)T^\epsilon$ underalgebraic cells from a finite number (depending on ϵ) of underalgebraic cell families.

If one takes the union of positive-dimensional cells from these families, one obtains a definable family, and this proves Theorem 9.14. □

Theorems 2.10 and 2.13 follow from Theorem 9.14. See also the treatment of parameterization and point-counting in [60], and a counting result in o-minimal structures enriched by a dense set (and so no longer o-minimal) in [203].

For an application of o-minimality in lattice-point counting, see [43]. For an application of o-minimality and point-counting in combinatorial geometry, see [219]. For (conjectural) connections with hyperbolic three-manifolds, see [292].

Counting Semi-rational Points

In the above proof, the bounded degree points in Z are controlled via the rational points in the set X that projects to it. More generally, for a definable set $Z \subset \mathbb{R}^m \times \mathbb{R}^n$, one can consider *semi-rational points*; that is, points that are rational only on the $\mathbb{R}^m$ coordinates.

9.15 Definition   For a definable set $Z \subset \mathbb{R}^m \times \mathbb{R}^n$, we set

$$Z^{\text{semi}}(k, T) = \left\{ (w, z) \in Z : H_k^{\text{poly}}(w) \le T \right\}.$$

In the following, we write π_1, π_2 for the projections of $\mathbb{R}^m \times \mathbb{R}^n$ to $\mathbb{R}^m, \mathbb{R}^n$, respectively.

9.16 Theorem ([264, Corollary 7.2])   *Let $Z \subset \mathbb{R}^\ell \times \mathbb{R}^m \times \mathbb{R}^n$ be a definable family of sets $Z_y \subset \mathbb{R}^m \times \mathbb{R}^n$, $k \ge 1$ and $\epsilon > 0$. There exists a constant $c(Z, k, \epsilon)$ with the following property. If $y \in \mathbb{R}^\ell$ and $\Sigma \subset (Z_y)^{\text{semi}}(k, T)$ with*

$$\#\pi_2(\Sigma) > cT^\epsilon,$$

then there exists a continuous definable function

$$\beta : [0, 1] \to Z_y$$

such that the following properties hold.

  (i) *The composition $\pi_1 \circ \beta : [0, 1] \to \mathbb{R}^m$ is semi-algebraic and its restriction to $(0, 1)$ is real analytic.*
 (ii) *The composition $\pi_2 \circ \beta : [0, 1] \to \mathbb{R}^n$ is non-constant.*
(iii) *We have $\pi_2(\beta(0)) \in \pi_2(\Sigma)$.*
 (iv) *If the o-minimal structure admits analytic cell decomposition, then $\beta|_{(0,1)}$ is real analytic.*

See also [118].

Diophantine Approximation in O-Minimal Structures

Habegger [260] adapts the counting methods to estimate rational points *near* a definable set. For an integer $k \ge 1$ and real $T \ge 1$ set

$$Q^n(k, T) = \left\{ q \in (\overline{\mathbb{Q}} \cap \mathbb{R})^n : H(q) \le T \text{ and } [\mathbb{Q}(q) : \mathbb{Q}] \le k \right\}.$$

For $Z \subset \mathbb{R}^n$ and $\epsilon > 0$ set,

$$\mathcal{N}(Z, \epsilon) = \left\{ y \in \mathbb{R}^n : \exists z \in Z \text{ with } |y - z| \leq \epsilon \right\}.$$

9.17 Theorem ([260, Theorem 2]) *Let $Z \subset \mathbb{R}^n$ be closed and definable in a polynomially bounded o-minimal structure. Let $k \geq 1$ be an integer and $\epsilon > 0$. There exists $c(Z, k, \epsilon) \geq 1, \lambda = \lambda(Z, k, \epsilon) > 0$, and $\theta = \theta(Z, k, \epsilon) \in [0, 1]$ such that for all $T \geq 1$ we have*

$$\#\left\{ q \in \mathbb{Q}^n(k, T) : \exists z \in Z \quad \text{with} \quad |z - q| \leq T^{-\lambda} \quad \text{and} \right.$$
$$\left. q \notin \mathcal{N}\left(Z^{\mathrm{alg}}, |z - q|^\theta \right) \right\} \leq cT^\epsilon.$$

The restriction to closed sets is necessary as algebraic subsets in the boundary of Z could lead to many nearby rational points. The assumption of polynomially bounded structures is needed for a similar reason as otherwise Z could approach a semi-algebraic set very closely. For similar reasons, the constant is not uniform in families. However, the restriction of exp to the unit circle is so definable, and an application is given to the question of how small a non-vanishing sum of a fixed number of roots of unity can be (see also [261]).

Non-Archimedean Point-Counting

The analogue of the Counting Theorem for p-adic subanalytic sets is given by Cluckers–Comte–Loeser [142], for which they establish a C^r-parameterization theorem (along with a notion of parmeterization in the totally disconnected topology that ensures that functions are well-approximated by their Taylor polynomials).

9.18 Theorem ([142]) *Let $Z \subset \mathbb{Q}_p^n$ be a subanalytic set of dimension $k < n$ and $\epsilon > 0$. Then there exists a positive constant $C = C(Z, \epsilon)$ and a semi-algebraic set $W = W(Z, \epsilon) \subset \mathbb{Q}_p^n$ such that $W \cap Z \subset Z^{\mathrm{alg}}$ and, for $T \geq 1$,*

$$N(Z - W, T) \leq CT^\epsilon.$$

The result is proved uniformly in families and for points of bounded degree. They also give counting results over the base field $\mathbb{C}((t))$. See also [480, 522] for results over function fields. Uniformity over function fields of varying characteristic is pursued in [143], also giving a uniform non-archimedean generalization of Theorem 2.10. The proofs rely on uniform C^r-parameterization in these settings.

For results on Wilkie's conjecture (10.1) in non-archimedean settings, see [72].

10

Better Bounds

The Counting Theorem holds in an arbitrary o-minimal expansion of the real field (or even of a real closed field). Its main drawbacks are that it is ineffective (for any given ϵ), and, in general, gives no control on the dependence of constants on ϵ. One cannot hope to address either of these in full generality: the examples in [421], mentioned in Remark 2.11.5, are of analytic curves for which the growth of $c(Z, \epsilon)$ with ϵ can be made arbitrarily fast. Thus, to have the possibility of stronger and effective bounds one must restrict to more specific classes of sets, for example by restricting in the first instance to specific o-minimal structures.

Wilkie (see [439, Conjecture 1.11]) conjectured a poly-logarithmic bound on the counting function for definable sets in $\mathbb{R}_{\exp}$. This conjecture has guided a number of partial results for definable sets in $\mathbb{R}_{\exp}$, and more generally in $\mathbb{R}_{\mathrm{pfaff}}$ (and its reducts), and other settings where the geometry of definable sets is explicitly controlled. Some results are further limited to sets of low dimension.

For diophantine applications one can sometimes restrict to rather particular kinds of sets, obtaining strong bounds.

The discussion below is divided between quality of bounds, methods of parameterization, and effectivity, but there is considerable overlap as the sets amenable to better quality bounds are often, for the same reasons, amenable to effective bounds.

Quality of Bounds

10.1 Wilkie's Conjecture   *For set Z definable in $\mathbb{R}_{\exp}$ there are constants c, γ depending on Z such that, for $T \geq e$,*

$$N(Z^{\mathrm{trans}}, T) \leq c \left(\log T \right)^{\gamma}.$$

Note: A proof of this conjecture has been announced ([558]).

We describe a number of results towards and around this conjecture. Explicit strong bounds ([223, 303]) for the number of connected components of semi-Pfaffian sets (sets defined by equalities and inequalities of Pfaffian functions) play a crucial role in many of them. It is then suggestive to extend Wilkie's conjecture to $\mathbb{R}_{\text{pfaff}}$ (as in [423]). However, in the absence of model completeness, the definable sets in $\mathbb{R}_{\text{pfaff}}$ may be quite a bit wilder than $\mathbb{R}_{\text{exp}}$, and this extension might be less natural than it seems.

Several different approaches to parameterization also play a role, and are taken up in more detail in the next chapter.

Bounds of this form seem to be what is qualitatively required to prove results in transcendental number theory, and guided by that it was further conjectured (see [426]) that a similar bound holds for points definable over a number-field with the exponent being independent of the number-field. Also, that a similar bound (with constants depending on k) holds for $N(k, Z^{\text{trans}}, T)$.

10.2 Definition   Let K be a number-field and $Z \subset \mathbb{R}^n$. Define

$$Z(K,T) = \{z \in Z \cap K^n : H(z) \leq T\}, \quad N(K,Z,T) = \#Z(K,T).$$

Several results establish a bound

$$N(K, Z^{\text{trans}}, T) \leq c(\log T)^{\gamma},$$

with $c = c(K, Z)$ and $\gamma = \gamma(Z)$ independent of K.

Schmidt [475] shows that some of these results can be further refined to give

$$c(K, Z) = C[K : \mathbb{Q}]^{\delta}$$

for C, δ depending only on Z (or even perhaps only on some complexity measure of Z). Applying these bounds to the graphs of the uniformizing maps of some one-complex-dimensional group varieties yields lower bounds for the Galois orbits of torsion points. Such points lead to algebraic points on the graph. The point-counting approach to Galois orbit lower bounds is described further in Chapter 11.

10.3 Definition   A function $f : U \to \mathbb{R}$, definable in $\mathbb{R}_{\text{pfaff}}$, with $U \subset \mathbb{R}^m$, is said to be *implicitly defined* (by Pfaffian functions) if there exists $n \geq 1$, Pfaffian functions $p_1, \ldots, p_n : \mathbb{R}^{m+n} \to \mathbb{R}$, and definable smooth functions $f_1, \ldots, f_n : U \to \mathbb{R}$ such that $f_1 = f$ and

$$p_1(x, f_1(x), \ldots, f_n(x)) = \cdots = p_n(x, f_1(x), \ldots, f_n(x)) = 0,$$

$$\det\left(\frac{\partial(p_1, \ldots, p_n)}{\partial(x_{m+1}, \ldots, x_{m+n})}\right)(x, f_1(x), \ldots, f_n(x)) \neq 0$$

for all $x = (x_1, \ldots, x_m) \in U$. If the functions $p_1, \ldots, p_n$ have a common Pfaffian chain of length r and degree (α, β), then f has an implicit definition of *complexity* (n, r, α, β).

10.4 Theorem ([298, 423, 475]) *Let $Z \subset \mathbb{R}^2$ be the graph of a function f implicitly defined by Pfaffian functions and suppose Z is not semi-algebraic. Then, for $T \geq e$,*

$$N(K, Z, T) \leq c[K : \mathbb{Q}]^\delta (\log T)^\gamma,$$

where c, δ, γ depend only on the complexity of the implicit definition of f.

Jones–Thomas [298] extend this to affirm Wilkie's conjecture for the graphs of univariate functions existentially definable in $\mathbb{R}_{pfaff}$ (and for points over number-fields). In view of the model-completeness of $\mathbb{R}_{exp}$, this affirms Wilkie's conjecture for all one-dimensional sets $Z \subset \mathbb{R}^n$ definable in $\mathbb{R}_{exp}$, a result also obtained by Butler [113].

Some results for unrestricted Pfaffian sets of dimension two are in [113, 426]. For example, [426] considers the surface

$$Z = \{(x, y, z) \in (0, \infty)^3 : \log x \, \log y = \log z\}.$$

Using Ax–Schanuel (see Chapter 13) it is shown in [426, §4] that

$$Z^{alg} = L_x \cup L_y \cup \bigcup_{q \in \mathbb{Q}^\times} \Gamma_{x,q} \cup \bigcup_{q \in \mathbb{Q}^\times} \Gamma_{y,q},$$

where $L_x = \{(x, 1, 1) : x > 0\}, L_y = \{(1, y, 1) : y > 0\}$ and, for $q \in \mathbb{Q}^\times$,

$$\Gamma_{x,q} = \{(x, e^q, z) : z = x^q, x > 0\}, \quad \Gamma_{y,q} = \{(e^q, y, z) : z = y^q, y > 0\}.$$

However, the sets $\Gamma_{x,q}, \Gamma_{y,q}$ have no algebraic points (by the transcendence of e, due to Hermite). Assuming Schanuel's conjecture, Z^{trans} has no algebraic points.

In [426] the following is proved affirming Wilkie's conjecture for the surface Z. For related results, see [113].

10.5 Theorem ([426, Theorem 1.6]) *Let $Z = \{(x, y, z) \in (0, \infty)^3 : \log x \, \log y = \log z\}$. Then, for a suitable constant c, with $T \geq e$,*

$$N(Z^{trans}, T) \leq c (\log T)^{45}.$$

To get such a poly-logarithmic bound using r-parameterizations one needs to allow the order r to increase with T, roughly as some positive power of $\log T$. This is the approach used in some results in [146] stated in Theorem 10.8. The proof of Theorem 10.5 uses instead a *mild parameterization* by C^∞ functions with successive derivatives satisfying suitable bounds (see Definition 10.10); see [520] for an improvement.

If one restricts to $\mathbb{R}_{\text{respfaff}}$, then every definable set has a mild parameterization ([294], indeed by real analytic functions). This leads to a poly-logarithmic bound for the counting function of the graph of an arbitrary $\mathbb{R}_{\text{respfaff}}$-definable univariate function. Mild parameterization is used by Jones–Thomas in [298] to affirm Wilkie's conjecture for surfaces definable in $\mathbb{R}_{\text{respfaff}}$, though in general the exponent and the constant are not uniform in families.

10.6 Theorem (Jones–Thomas [298]) *Let $Z \subset \mathbb{R}^n$ be definable in $\mathbb{R}_{\text{respfaff}}$ with* $\dim Z = 2$, *and $K \subset \mathbb{R}$ a number-field. Then there are constants $c(X, K), \gamma(X)$ such that, for $T \geq e$,*

$$N(K, Z^{\text{trans}}, T) \leq c\left(\log T\right)^{\gamma}.$$

See [295] for Pfaffian definitions with uniform complexity for the uniformizing maps of elliptic curves (which are real surfaces).

These methods do not seem to go beyond surfaces, as it does not seem possible to control the parameterizations of the resulting hypersurface intersections. If the starting set Z has real dimension three, then those intersections are surfaces and must be controlled for hypersurfaces of varying degrees. These do not form a definable family. In all the results above, the intersection is a curve and can be dealt with by more direct means in a Pfaffian context.

Some further improvements are obtained in [146], essentially showing that the constants in Theorem 10.6 can be made uniform for certain definable families of Pfaffian surfaces (i.e. graphs of Pfaffian functions of two variables). We do not state the results, but we describe below the two parameterization techniques introduced in [146], which are applicable more broadly to $\mathbb{R}_{\text{an}}$ (one of them even in $\mathbb{R}_{\text{an}}^{\text{pow}}$). The parameterizations apply to sets in arbitrary dimension, but for the same reasons as above the diophantine applications are limited to surfaces in the Pfaffian setting.

However, if one restricts further to the structure of restricted elementary functions (see Definition 8.33), then Binyamini–Novikov [76] affirm Wilkie's conjecture for definable sets of all dimensions.

10.7 Theorem ([76]) *Let $Z \subset \mathbb{R}^n$ be definable in $\mathbb{R}_{\text{RE}}$. Then, for $T \geq e$,*

$$N(K, Z^{\text{trans}}, T) \leq c\left(\log T\right)^{\gamma}$$

for suitable constants $c = c(Z, K), \gamma = \gamma(Z)$.

They prove also a poly-logarithmic estimate for all algebraic number of a fixed degree (with exponent depending on the degree).

The parameterization is here obtained with ideas from complex analytic geometry using Weierstrass polydiscs (which are closely related to the *quasi-parameterization* of [146]; see Theorem 10.14), but also requires Gabrielov–Vorobjov bounds. Thus, the need to work in the more restrictive structure of Pfaffian functions whose complexification remains Pfaffian.

Another approach to point-counting is introduced by Binyamini [71]. The results apply to foliations defined over number-fields, and are applicable in the Shimura variety settings. The results are poly-logarithmic in height and also in degree, which has led to a number of (effective) diophantine applications, in particular to obtaining lower bounds for Galois orbits under suitable conditions (see Chapter 11).

How Should One Parameterize a Definable Set?

There are many contexts and reasons for parameterizing or uniformizing a set. See, for example, [77, 539]. Here we consider this issue only from the standpoint of point-counting.

The Counting Theorem is proved using r-parameterizations, which always exist for a definable set in an o-minimal structure. In pursuing improved bounds (in quality or effectivity) there are two main issues to face. One is to control how the complexity of the r-parameterization (i.e. the number of parameterizing functions required) increases with r. The second issue is to control the parameterizations of the resulting hypersurface intersections (including how they depend on the order of parameterization for these) and is discussed further below.

To get poly-logarithmic point-counting bounds using r-parameterizations, one needs, as mentioned, to take r increasing with T, roughly as some positive power of $\log T$, with polynomial growth in the complexity. This is achieved by one of the parameterization methods in Cluckers–Pila–Wilkie [146], for sets definable in $\mathbb{R}_{an}^{pow}$ (see Theorem 10.14 for another method).

10.8 Theorem (The Uniform C^r-Parameterization Theorem; [146, Theorem 2.1.3]) *Let $Z \subset Y \times (0,1)^n$ be an $\mathbb{R}_{an}^{pow}$-definable family of sets Z_y of dimension k for $y \in Y$. For each positive integer r there exists a finite set Φ of (definable) families of analytic maps*

$$\phi_{r,t} : (0,1)^k \to Z_y, \quad y \in Y,$$

whose C^r-norms are bounded by 1 and whose ranges cover Z_y, and with

$$\#\Phi \le cr^\gamma$$

for some constants $c = c(Z), \gamma = \gamma(Z)$.

The polynomial growth of #Φ applies also in suitable reducts (see [146, 2.2.1]), including $\mathbb{R}_{alg}$, where it partially answers a question in [112, below 3.8], see also [538, Remark 3]. Very precise control of C^r-parameterizations in the semi-algebraic case is obtained in [77] (with applications in the original context of dynamics). It is also shown in [77, Theorem 2] that for families in $\mathbb{R}_{an}$ one may take $\gamma(Z) = k$ in the theorem. (In $\mathbb{R}_{an}^{pow}$, one may take $\gamma(Z) = k^2$; see [519].)

Theorem 10.8 leads to the following result on the existence of suitable hypersurface intersections that contain the rational points up to a given height.

10.9 Theorem ([146, Theorem 2.3.1]) *Let Z be a family as in Theorem 10.8. Then there exist constants $c(Z), \gamma(Z)$ such that, for $T \geq e$ and $y \in Y$, $Z_y(\mathbb{Q},T)$ is contained in the union of at most $c(\log T)^\gamma$ real algebraic hypersurfaces (possibly reducible) of degree at most*

$$d = [(\log T)^{k/(n-k)}].$$

Here, $[c] \in \mathbb{Z}$ is the integer part of c, having $c \leq [c] < c + 1$.

A second approach is to parameterize the set by C^∞ functions with sufficient control on the sizes of *all* derivatives to achieve poly-logarithmic growth in point-counting. The requirements are met with a *mild parameterization*, introduced in [426], which is more general than parameterization by real analytic functions (essentially, it is parameterization by Gevrey functions; see [237, 519]).

10.10 Definition A (J, A, C)-*mild parameterization* of a k-dimensional set $Z \subset (0,1)^n$ is a finite set $\Phi = \{\phi\}$, $\#\Phi = J$, of C^∞ maps $\phi : (0,1)^k \to (0,1)^n$, the union of whose images is Z, and with the property that the partial derivatives of the coordinate functions ϕ_i of each ϕ to all orders $\mu \in \mathbb{N}^k$ satisfy

$$|\partial^\mu \phi_i(z)| \leq \mu! \left(A|\mu|^C\right)^{|\mu|}.$$

A *mild parameterization* is a (J, A, C)-mild parameterization for some J, A, C.

If a definable set $Z \subset (0,1)^n$ of dimension k has a (J, A, C)-mild parameterization, then the conclusion of Theorem 10.9 holds with suitable $c(n, k, J, A, C)$, $\gamma(n, k, J, A, C)$ and the same d; see [426, Theorem 3.2].

10.11 Note It is tacitly assumed in the proof of Theorem 3.2 in [426] that $A \geq 1$. This does not affect the subsequent results where only the parameter C is used explicitly, but the precise form of that theorem is established only for $A \geq 1$.

This is the approach adopted in [113, 426] for some specific unrestricted Pfaffian sets admitting a mild parameterization. While some o-minimal structures do not admit mild parameterization (see [500]), Jones–Miller–Thomas [294] show that all definable sets in $\mathbb{R}_{\text{respfaff}}$ admit such a parmeterization with $C = 0$; that is, by real analytic maps. See [521] for results on families of bounded sets definable in $\mathbb{R}_{\text{an}}^{\text{pow}}$. Yomdin [538, Proposition 3.3] shows that suitable analytic parameterizations (with a fixed A) cannot be uniform in parameters, even for curves. This is done using the example $xy = \epsilon$ in $(0, 1)^2$ (a C^r-parameterization exercise proposed in [248, right after 3.3]). However, uniformity can be achieved with $(A, 2)$-mild maps; see also [520].

10.12 Theorem (Binyamini–Novikov [77, Theorem 3]) *Let $Z \subset Y \times (0, 1)^n$ be an $\mathbb{R}_{\text{an}}$-definable family of sets Z_y of dimension k for $y \in Y$. There exist $J = J(Z), A = A(Z)$ such that every fibre Z_y has a $(J, A, 2)$-parameterization.*

A third approach uses a more general *quasi-parameterization* related to Weierstrass preparation that can be carried out in $\mathbb{R}_{\text{an}}$ (or suitable reducts).

10.13 Definition   Let $R, K > 0$ and $\Delta(R)$ be the open unit ball in $\mathbb{C}$ of radius R. A definable family $F = \{F_y : y \in Y\}$, where $Y \subset \mathbb{R}^m$ is a definable set, is called an (R, k, K)-family if, for each $y \in Y$, the function

$$F_y \colon \Delta(R)^k \to \mathbb{C}$$

is holomorphic and $|F_y(z)| \leq K$ for all $z \in \Delta(R)^k$.

10.14 Theorem (The Quasi-Parameterization Theorem; [146, Theorem 2.2.3]) *Let $Z = \{Z_y : y \in Y\}$ be an $\mathbb{R}_{\text{an}}$-definable family of subsets $Z_y \subset [-1, 1]^n$, with $\dim Z_y \leq k$ for all $y \in Y$. Then there exist $R > 1, K > 0$, a positive integer d, and an $(R, k + 1, K)$-family $F = \{F_x, x \in X\}$ such that each F_x is a monic polynomial of degree d in its first variable and, for all $y \in Y$, there exists $x \in X$ such that*

$$Z_y \subset \left\{ z = (z_1, \ldots, z_n) \in [-1, 1]^n : \exists w \in [-1, 1]^m \bigwedge_{i=1}^{n} F_x(z_i, w) = 0 \right\}.$$

A similar parameterization, by means of Weierstrass polydiscs, is employed in [75] to give a complex-analytic proof of the Counting Theorem for globally subanalytic sets, and in [76] to prove Wilkie's conjecture for restricted elementary sets. Note that in [75, Theorem 2] the set and the hypersurfaces are complex.

10.15 Theorem ([146, Theorem 2.3.2]; cf. [76, Theorem 2]) *Let $Z = \{Z_y : y \in Y\}$ be an $\mathbb{R}_{\text{an}}$-definable family of sets in $[-1, 1]^n$, with $\dim Z_y \leq k$. Then there is a constant $c = c(Z)$ such that if $T \geq e$ and $y \in Y$, then $Z_y(\mathbb{Q}, T)$ is* *contained in the union of at most c(Z) real algebraic hypersurfaces of degree at most*

$$d = \left[\left(\log T\right)^{k/(n-k)}\right].$$

Another method of parameterization is introduced in Binyamini–Novikov [77]. They show that the obstruction to uniform analytic parameterization comes essentially from hyperbolic geometry, and can be remedied by admitting more general domains for the parameterizing functions. These *complex cells* clarify many of the issues and enable the results cited above on parameterization of subanalytic and semi-algebraic sets, with sharpenings in the applications to point-counting on algebraic varieties already mentioned in Chapter 2.

It would take us too far afield to describe this theory, but we note the following further progress from the viewpoint of diophantine applications around the André–Oort–Zilber–Pink conjectures.

The sets where counting takes place are not globally subanalytic, and they are not in general Pfaffian ([220]). However, at least in some key cases, they are *log sets*, as defined in [77, Definition 4]: the (principal value of) logarithms of globally subanalytic sets (in $\mathbb{C}^n$, identified with $\mathbb{R}^{2n}$). For example, the j-function factors through the q-expansion (see [77, B.3.2]). The counting result ([77, Proposition 5]) is a promising approach to proving poly-logarithmic bounds in such settings. See further discussion in [77, 1.4.5].

The second issue to be faced is the iterative use of parameterizations of lower-dimensional families of intersections. This seems difficult to control in general (a sufficient conjecture is framed in [426] for mild parameterizations). It is shown in [76] that this can be controlled in $\mathbb{R}_{RE}$.

What could one hope for? One cannot parameterize a set with fewer maps than the number of connected components, and the number of connected components in a hypersurface intersection grows at least polynomially with the degree of the hypersurface. Gabrielov–Vorobjov [223] bounds achieve a polynomial upper bound on the connected components of such intersections, and it is plausible that one could achieve control on derivatives with a parameterization of similar complexity.

Such parameterizations for general definable sets in $\mathbb{R}_{exp}$, let alone the larger structures required for the diophantine uniformizations, seem to be still some way off.

Effective Point-Counting

The Counting Theorem is ineffective as it relies on o-minimality to bound various quantities that arise. O-minimality is a qualitative concept (certain sets are simply assumed to be finite). However, some effective results have been obtained for special kinds of definable sets.

Jones–Thomas [300] prove an effective version of the Counting Theorem for surfaces that are the graphs of unrestricted Pfaffian functions.

10.16 Theorem (Jones–Thomas [300])   *Let $f : \mathbb{R}^2 \to \mathbb{R}$ be a Pfaffian function of order r and degree (α, β) with graph Z, and let $\epsilon > 0$. Then there exists an effectively computable constant $c = c(r, \alpha, \beta, \epsilon)$ such that, for all $T \geq 1$,*

$$N(Z^{\mathrm{trans}}, T) \leq cT^\epsilon.$$

The proof uses a family of restricted Pfaffian functions and leverages the uniformity in families to get the result for unrestricted Pfaffian functions. Note also that uniformity over Pfaffian functions of a given complexity may be a wider uniformity than over definable families.

Binyamini [69] gives effective point-counting results for Noetherian functions on bounded domains. Noetherian functions satisfy algebraic differential equations, but without the triangularity condition required for Pfaffian functions.

10.17 Definition   Let $U \subset \mathbb{R}^n$ be a bounded domain with closure $\overline{U}$.

1. A finite tuple of real analytic functions $\phi = (\phi_1, \ldots, \phi_r) : \overline{U} \to \mathbb{R}^r$ is called a *(real) Noetherian chain* if there are polynomials

$$P_{i,j} \in \mathbb{R}[X_1, \ldots, X_n, Y_1, \ldots, Y_r]$$

   such that

$$\frac{\partial \phi_i}{\partial x_j} = P_{i,j}(x, \phi), \quad i = 1, \ldots, r, \quad j = 1, \ldots, n.$$

   The *order* of the chain is r and the *degree* of the chain is $\alpha = \max_{i,j} \deg P_{i,j}$.
2. If $Q \in \mathbb{R}[X, Y]$, then $f = Q(x, \phi)$ is a *Noetherian function of degree β*.
3. The set of common zeros of a collection of Noetherian functions is called a *real Noetherian variety*. A set defined by a finite system of Noetherian equalities or inequalities is called a *basic semi-Noetherian set*, and a finite union of such sets is called a *semi-Noetherian set*.
4. A semi-Noetherian set has a degree, which bounds the discrete parameters, and a *size* given by the sizes of the coordinate functions and the ϕ_i on U and the size of the coefficients of the polynomials involved.

10.18 Theorem ([69])   *Let $Z \subset U$ be a semi-Noetherian set and $\epsilon > 0$. Then,*

$$N(Z^{\mathrm{trans}}, T) \leq cT^\epsilon,$$

where the constant can be explicitly estimated in terms of ϵ and the Noetherian degrees and sizes.

The results of [71] are also effective.

Noetherian versus Pfaffian versus Definable Functions

One does not have Pfaffian-type finiteness statements for Noetherian sets. For example, the sine function is Noetherian and has infinitely many isolated zeros. One can bound local multiplicities and, in some cases, get bounds for the number of zeros in a compact region. These issues are close to transcendence theory.

A prime example is the uniform bound of Tijdeman [501] for zeros of exponential polynomials.

10.19 Theorem ([501])   *Let $\omega_i \in \mathbb{C}$, and $P_i \in \mathbb{C}[Z], i = 1,\ldots,m$, with $\deg P_i = d_i$. Suppose*

$$f(z) = \sum_{i=1}^{m} P_i(z)e^{\omega_i z}$$

is not identically zero. Let $n = \sum_i d_i$ and $\Delta = \max_i(|\omega_i|)$. Then the number of zeros of f in a ball of radius R around $z_0 \in \mathbb{C}$ is at most

$$3(n + m - 1) + 4R\Delta.$$

Now $\exp(z)$ is definable on horizontal strips, so a uniform finiteness holds on strips of width R rather than discs if one takes real frequencies ω_i with $|\omega_i| \leq 1$ and fixed m and d_i. If one takes $\omega_i \in \mathbb{N}$, then the problem reduces to estimating zeros of $Q[z, e^z]$ in a strip for $Q \in \mathbb{C}[X, Y]$ of degree d. Here explicit bounds; polynomial in d and R follow from Gabrielov–Vorobjov bounds due to the fact that e^z is Pfaffian on horizontal strips considered as a map $\mathbb{R}^2 \to \mathbb{R}^2$.

The j-function is Noetherian but not Pfaffian ([220]). In a compact domain, explicit polynomial bounds are obtained in [68] for the number of zeros of a polynomial of degree d in z and $j(z)$; in fact, the results allow for several $j(r_i(z)), j'(r_i(z)), j''(r_i(z))$, where $r_i(z)$ are rational functions with algebraic coefficients. By o-minimality, given d, a uniform bound holds for the number of zeros of $F(z, j(z))$, where $F \in \mathbb{C}[Z, W]$ is of degree d, in the classical fundamental domain. As the fundamental domain is not compact, such a bound does not follow from the aforementioned results.

10.20 Theorem (Armitage [12])   *Let $P \in \mathbb{C}[X, Y]$ of degree d. Then $P(z, j(z))$ has at most $2^{68} d^{10}$ zeros in the fundamental domain.*

11

Point-Counting and Galois Orbit Bounds

In this chapter, we describe how point-counting bounds with sufficiently strong dependencies on height and degree, in conjunction with height upper bounds, can be used to prove lower bounds for the Galois orbits of special points.

This is initiated in [475], using the results stated in Theorem 10.4 above, to obtain Galois orbit lower bounds for roots of unity and for torsion points of ellliptic curves. While weaker than bounds known by other means, they are of the right quality for applications of the point-counting strategy. However, the counting results in [475] are for sets with either Pfaffian definitions or mild parameterizations, and are not generally applicable for Shimura varieties.

The point-counting method ([71]) for foliations defined over number-fields (mentioned after Theorem 10.7) is applicable to general Shimura varieties and leads to bounds of the right shape for the deduction of Galois orbit lower bounds. Binyamini–Schmidt–Yafaev [79] carry out this approach, which we now sketch.

Let X be a Shimura variety with uniformization $u : \Omega \to X$ invariant under Γ and semi-algebraic fundamental domain $F \subset \Omega$. Let L be a number-field over which X is defined. Let

$$Z \subset \Omega \times X$$

be the restriction to F of the graph of u:

$$Z = \{(z,x) : x \in F, x = u(z)\}.$$

The algebraic part of Z is empty. It is convenient to define a variant of the set $Z(k,T)$ building in the logarithmic height dependence (below, h is the logarithmic height; see just before Definition 2.12):

$$Z[f,h] = \{(z,x) \in Z : [L(z,x) : L] \le f, h(z,x) \le h\}.$$

11.1 Theorem ([79, Theorem 3]) *With the notation as above,*

$$\#Z[f, h] \leq C(fh)^c$$

for some constants C, c (depending on the data X, F, u, L).

A special point $x \in X$ has algebraic pre-image $z \in F$, and thus gives rise to an algebraic point $(z, x) \in Z$.

Let us specialize to the case of the modular function $j \colon \mathbb{H} \to Y(1)$. An imaginary quadratic order, of discriminant D, gives rise to $h(D)$ (the class number) special points (singular moduli). Let σ be one of these special points, with complexity $\Delta(\sigma) = |D|$ (see Definition 4.5.3). The pre-image of σ in the fundamental domain is a quadratic point of multiplicative height at most $2|D| = 2\Delta(\sigma)$ (see Ingredient 5.9), while the logarithmic height of σ satisfies

$$h(\sigma) \leq c(\epsilon)\Delta(\sigma)^\epsilon$$

(see Theorem 21.12). We have $|D|^{1/2-\epsilon} \ll_\epsilon h(D)$ and the counting bound gives $h(D) \ll_\epsilon f^c |D|^\epsilon$, giving the following.

11.2 Corollary *For some $c > 0$ we have*

$$f = [\mathbb{Q}(s) : \mathbb{Q}] \gg \Delta(\sigma)^c.$$

In the modular case it is known that the points are all conjugate, giving a better exponent. However, the same deduction can be made in the general case. The complexity $\Delta(x)$ of a special point $x \in X$ is measured by the discriminant of the splitting field and index data of the Mumford–Tate group $\mathbf{T}$ of x (a torus; see the discussion before Theorem 6.13), a power of which bounds the multiplicative height $H(z)$ of the pre-images (Theorem 6.13). Associated with $\mathbf{T}$ is a zero-dimensional special subvariety $S(x)$. The points of $S(x)$ are all special points of the same discriminant $\Delta(x)$ and all have the same degree over L.

11.3 Proposition ([79, Proposition 6]) *The number of special points in the associated zero-dimensional special subvariety is equal to the size of the class group of $\mathbf{T}$.*

The size of the class group of $\mathbf{T}$ is bounded below by a positive power of $\Delta(x)$, as established independently in [513] and [504].

11.4 Theorem ([79, Theorem 1]) *Assume that for special points $x \in X$, one has*

$$h(x) \ll_{X,\epsilon} \Delta(x)^\epsilon$$

for every $\epsilon > 0$. Then,

$$[L(x) : L] \gg_X \Delta(x)^c$$

for some positive constant $c(X)$.

Under the height hypothesis, one says that the heights of special points are *discriminant negligible*.

12

Complex Analysis in an O-Minimal Structure

The familiar identification of $\mathbb{C}$ with $\mathbb{R}^2$ via real and imaginary parts, in which the field operations are expressible algebraically via their real counterparts, shows that the complex field is interpretable in the real field. More generally, if R is a real closed field, then $K = R(\sqrt{-1})$ is algebraically closed; however, an algebraically closed field, such as $\mathbb{C}$, contains many different real closed subfields R, including non-archimedean ones.

In a series of papers (including [409, 410, 412]; see also [414]), Peterzil–Starchenko develop complex analysis in the setting of an o-minimal expansion of a real closed field. In general the underlying real field need not be archimedean, so one cannot base the development on convergent power series or integration.

Instead the development is via topological analysis, in particular winding numbers. The basic results of the classical theory are recovered, but now available in more general settings. It also yields strong versions of some classical results, such as Removable Singularities, Chow's Theorem, and Remmert–Stein.

These definable versions, roughly speaking, replace compactness assumptions by definability assumptions. The results are very flexible and powerful, and are new even in the classical setting. The purpose of this chapter is to state the basic definitions and results, including the Definable Chow Theorem.

Let R be a real closed field and $K = R(\sqrt{-1})$ be the algebraic closure of R. Choosing $\sqrt{-1}$, one may identify K with R^2, and the field operations are semi-algebraic. One has the order topology on R, and the product topology on K. This gives a natural notion of limits of functions.

12.1 Definition

1. Let $U \subset K$ be an open set. A function $f : U \to K$ is *K-differentiable* at

$z_0 \in U$ if the limit

$$\lim_{z \to z_0} \frac{f(z) - f(z_0)}{z - z_0}$$

exists in K. In that case, the limit is called the *K-derivative* of f at z_0 and denoted $f'(z_0)$. If f is K-differentiable at every $z \in U$, then it is called *K-holomorphic* on U.

2. Let $U \subset K^n$ be open and $f : U \to K$ continuous. Then f is called *K-holomorphic on U* if it is K-differentiable separately in each variable.

While the topology on R may not be locally compact or connected, tameness can be restored by restricting to functions that are definable in an o-minimal expansion $\mathcal{R} = (R, <, +, \times, \ldots)$. As observed in [409], R^n is definably connected (see [189, §1, 3.5]) and locally definably compact (see [409]: a definable set $X \subset R^n$ is called *definably compact* if every definable function $f : (a, b) \to X$ has a limit point in X as $t \to b$). With this restriction, the basic results about complex functions may be recovered.

Now no complex entire non-polynomial function can be definable in an o-minimal structure, by the Big Picard Theorem. More generally (so not by invoking Big Picard), a definable K-meromorphic function on K^n is a rational function ([410, Theorem 1.2.2]; see below). At the other extreme, every meromorphic function is locally definable in $\mathbb{R}_{an}$. In between, as we have already seen, the uniformizing maps of all mixed Shimura varieties are definable (in $\mathbb{R}_{an\,exp}$) on suitable fundamental domains.

In the following, $\Delta \subset K^2$ denotes the open unit disc and $\overline{\Delta}$ its closure.

12.2 Theorem ([414, Theorem 3.2]) *Let $U \subset K$ be open and $f : U \to K$ definable K-holomorphic.*

1. (Maximum Principle) *If $f : \overline{\Delta} \to K$ is a definable continuous function that is K-holomorphic on Δ, then $|f|$ attains its maximum on the boundary.*
2. (Open mapping theorem) *If $U \subset K$ is open and definable, and $f : U \to K$ is definable and K-holomorphic and non-constant, then f is an open map.*
3. (Infinite differentiability) *If $U \subset K$ is open and definable, and $f : U \to K$ is definable and K-holomorphic, then $f'(z)$ is also K-holomorphic on U.*
4. (Identity Theorem) *If $f : U \to K$ is definable and K-holomorphic in a neighbourhood of 0 and $f^{(k)}(0) = 0$ for all $k \in \mathbb{N}$, then f vanishes identically in a neighbourhood of 0.*

We state some theorems on removal of singularities.

12.3 Definition ([410, Definition 2.29]) Let $U \subset K^n$ be a definable open set. A definable partial function $f : U \to K$ is called *definably meromorphic* on U if

(i) the domain of f is of the form $U \backslash S$ for a definable S with $\dim_{\mathcal{R}} S < \dim_{\mathcal{R}} U$;

(ii) at every point $z_0 \in U$ there is a definable neighbourhood $U' \subset U$ of z_0 and definable K-holomorphic functions $g, h \colon U' \to K$ such that f and g/h (are defined and) agree on $U' \backslash S$.

12.4 Theorem (Removable Singularities Theorem; [410, Theorems 1.1 and 1.2]) *Let $U \subset K^n$ be a definable open set, $L \subset U$ a definable set, and $f \colon U \backslash L \to K$ a definable K-holomorphic function.*

(i) *If $\dim_{\mathbb{R}} L \leq 2n - 1$ and $f \colon U \to K$ is definable and continuous, then f is K-holomorphic on all U.*

(ii) *If $\dim_{\mathbb{R}} L = 2n - 2$, then f is definably meromorphic on U.*

(iii) *If $\dim_{\mathbb{R}} L = 2n - 2$ and $U = K^n$, then f is a rational function. In particular, every definably meromorphic function on K^n is a rational function.*

12.5 Theorem ([409, Lemma 2.41]) *Let $U \subset K$ be an open definable set, $z_0 \in U$, and let $f \colon U \backslash \{z_0\} \to K$ be a definable K-holomorphic function. Then:*

(i) *The limit of $f(z)$ as $z \to z_0$ exists (possibly equals ∞).*

(ii) *Either $f(z)$ or $1/f(z)$ can be extended, as a K-holomorphic function, to z_0.*

Here is the Definable Chow Theorem, stated in a slightly stronger form observed in [471] (see also [434, Lemma 4.3]).

12.6 Theorem (Definable Chow Theorem; [414, Theorem 4.5]; [471]) *Let Y be a quasi-projective algebraic variety and let $A \subset Y$ be definable, complex analytic, and closed in Y. Then A is algebraic.*

This includes the classical Chow Theorem in which Y is assumed projective, as the closed complex-analytic set A is then definable in $\mathbb{R}_{\mathrm{an}}$. For a non-archimedean version, see [401].

These results, in particular the Definable Chow Theorem, are crucial in proving the functional transcendence theorems we present in Part III, and are at the heart of some further applications in Hodge theory ([27, 29]). The latter develops an "o-minimal GAGA". See also [26, 28] and a non-model-theoretic approach in [103].

PART III

AX–SCHANUEL PROPERTIES

13

Schanuel's Conjecture and Ax–Schanuel

Schanuel's Conjecture

The following conjecture due to Schanuel is stated in [325, p. 30].

13.1 Conjecture (Schanuel's conjecture (SC))   *Let* $z_1, \ldots, z_n \in \mathbb{C}$. *Then*

$$\text{tr. deg.}_{\mathbb{Q}} \mathbb{Q}\left(z_1, \ldots, z_n, e^{z_1}, \ldots, e^{z_n}\right) \geq n$$

unless $z_1, \ldots, z_n$ *are linearly dependent over* $\mathbb{Q}$.

Equivalently, for any $z_1, \ldots, z_n \in \mathbb{C}$,

$$\text{tr.deg.}_{\mathbb{Q}} \mathbb{Q}\left(z_1, \ldots, z_n, e^{z_1}, \ldots, e^{z_n}\right) \geq \text{lin.dim.}_{\mathbb{Q}}\left(z_1, \ldots, z_n\right).$$

This is known when $n = 1$ (Hermite–Lindemann Theorem), and open in general for $n \geq 2$, though various (very) special cases are known.

Schanuel's conjecture asserts that exponentiation is "as transcendental as possible" and implies all the standard transcendence results and standard conjectures about exponentiation. We mention below a few results and conjectures whose functional analogues are relevant to our themes (Lindemann's Theorem was already stated as Theorem 3.9).

Lang [325, p. 31] observes: "From this statement, one would obtain most statements about the algebraic independence of values of e^t and $\log t$ which one feels to be true." This has been made precise in a certain sense by Zilber [551], who shows that SC is part of a conjecturally categorical axiomatization of a (suitable non-first-order) theory of complex exponentiation; see the discussion preceding Conjecture 18.13. For a more concrete formalization of this idea, see [305, Theorem 9.2].

13.2 Theorem (Lindemann(–Weierstrass) Theorem; see e.g. [325])   *Let* $a_1, \ldots, a_n \in \overline{\mathbb{Q}}$. *Then* $e^{a_1}, \ldots, e^{a_n}$ *are algebraically independent over* $\mathbb{Q}$ *unless* $a_1, \ldots, a_n$ *are linearly dependent over* $\mathbb{Q}$.

13.3 Theorem (Gelfond–Schneider Theorem; see e.g. [24, 325])   *Let $a \neq 0, 1$ and $b \notin \mathbb{Q}$ be algebraic numbers. Then a^b is transcendental.*

For example, $2^{\sqrt{2}}, 2^i$ are transcendental. Otherwise expressed, if $a = \alpha_1, \alpha_2 \in \overline{\mathbb{Q}}^\times$ are multiplicatively independent, then $\log \alpha_1, \log \alpha_2$ are linearly independent over $\overline{\mathbb{Q}}$. Baker's Theorem generalizes this.

13.4 Theorem (Baker's Theorem; see e.g. [24])   *Let $\alpha_1, \ldots, \alpha_n \in \overline{\mathbb{Q}}^\times$ be multiplicatively independent. Then, with any determination of their logarithms,*

$$1, \log \alpha_1, \ldots, \log \alpha_n$$

are linearly independent over $\overline{\mathbb{Q}}$.

Baker's Theorem is the strongest result known towards the conjectural algebraic independence of logarithms of multiplicatively independent algebraic numbers (which is implied by SC).

13.5 Conjecture (Algebraic independence of logarithms)   *Suppose algebraic numbers $a_1, \ldots, a_n$ are multiplicatively independent. Then any determination of their logarithms are algebraically independent over $\mathbb{Q}$.*

13.6 Periods

Another wide-ranging, and wide open, conjecture in transcendence theory is the Grothendieck Period conjecture (see [325, pp. 40–44]), see also the closely related conjecture on periods of Kontsevich–Zagier [314]. They can be viewed as generalizing the conjecture on algebraic independence of logarithms, since logarithms of algebraic numbers are periods. These conjectures (conjecturally) do not imply SC as, conjecturally, e is not a period. However, André's Generalized Period conjecture ([8, §23.4–5]) contains both. For an explication of its implications for 1-*motives*, which includes SC, elliptic versions, and implications for the modular function, see [47]. See [9, 23, 48] on the evolution of these conjectures. See also [153, 154] and [553] for a connection with o-minimality.

A particular implication of interest to us is the modular analogue of Lindemann's Theorem stated in Conjecture 5.7, and a version with derivatives in Conjecture 15.9.

Functional Transcendence

Both SC and the Period conjectures have functional analogues. For functional versions of the period conjecture see [5, 51], and [22], which proves a functional version of the Kontsevich–Zagier Period conjecture.

We concentrate on the functional analogues of SC as these arise in applications of point-counting to the Zilber–Pink conjecture to be presented below (we have already seen Ax–Lindemann in the context of André–Oort), though we consider an Ax-type result for periods in Chapter 17.

Ax–Schanuel

In 1971, Ax proved a functional analogue of Schanuel's conjecture that, according to Ax, had also been conjectured by Schanuel. This can be stated in various ways, the most general being a statement in a differential field (see Definition 13.8 and Theorem 13.10). However, it is equivalent (as shown in §13.12) to the following complex-analytic statement, in which dimension replaces transcendence degree. We start with this version. We consider exponentiation in several variables,

$$\exp \colon \mathbb{C}^n \to \left(\mathbb{C}^\times\right)^n$$

with graph

$$D \subset \mathbb{C}^n \times \left(\mathbb{C}^\times\right)^n.$$

13.7 Theorem (Complex Ax–Schanuel [20, 21]) *Suppose that $Y \subset \mathbb{C}^n \times (C^\times)^n$ is an algebraic variety and let U be a complex-analytically irreducible component of $D \cap Y$. Then,*

$$\dim U = \dim Y - n$$

unless the projection of U to $\mathbb{C}^n$ is contained in a translate of a proper rational subspace (i.e. a proper weakly special subvariety) of $\mathbb{C}^n$.

Note that

$$\dim Y - n = \dim D + \dim Y - \dim \left(\mathbb{C}^n \times \left(\mathbb{C}^\times\right)^n\right).$$

Thus, the right-hand side is the typical dimension one would expect U to have by adding codimensions. The dimension can never be less than this (see e.g. [337, III.4.6]), and the theorem says that the dimension is typical unless a proper weakly special subvariety intervenes.

Theorem 13.7 implies Ax–Lindemann for the exponential function (Theorem 3.8), and, at the other extreme, a statement about logarithms of algebraic functions. These are purely formal deductions that we leave to the next chapter.

Let $U \subset \mathbb{C}^n$ be complex analytic. It has the same dimension as the graph $D|_U$ of exponentiation on it. Let W be the Zariski closure of $D|_U$. Considering the coordinate functions $\bar{z}_i$ restricted to U and their exponentials, one finds that

$$\dim W = \text{tr. deg.}_\mathbb{C}\,\mathbb{C}\left(\bar{z}_1, \ldots, \bar{z}_n, e^{\bar{z}_1}, \ldots, e^{\bar{z}_n}\right) \geq n + \dim U$$

unless the $\bar{z}_i$ are linearly dependent over $\mathbb{Q}$ modulo $\mathbb{C}$ (see Definition 13.9), which displays the analogy with SC, and suggests the form in a differential field. The statement here is an inequality because U could be smaller than the component of $D \cap W$ that contains it.

Differential Field Formulation of Ax–Schanuel

13.8 Definition   Let K be a field.

1. A *derivation* on a field K is an additive map $D: K \to K$ satisfying the *Leibniz rule:* $Dxy = xDy + yDx$ for all $x, y \in K$.
2. A *differential field* (with several commuting derivations) is a field K equipped with a set $\Delta = \{D_1, \ldots, D_k\}$ of commuting derivations.
3. The *constant field* is the field $C = \bigcap_{D \in \Delta} \ker D$.
4. A differential field is called *finitely generated* if it is finitely generated as a differential field.

13.9 Definition   Let K be a field containing $\mathbb{Q}$ and C a subfield. We say that $x_1, \ldots, x_n \in K$ are *linearly independent over $\mathbb{Q}$ mod C* if there is no equality of the form

$$\sum_{i=1}^{n} q_i x_i = c,$$

where $q_i \in \mathbb{Q}$ are not all zero, and $c \in C$.

13.10 Theorem (Ax–Schanuel Theorem, differential field version [20])   *Let K be a differential field containing $\mathbb{Q}$ with derivations $D_1, \ldots, D_k$ and constant field C. Let $x_1, \ldots, x_n, y_1, \ldots, y_n \in K^\times$ with*

$$D_j y_i = y_i D_j x_i$$

for all $i = 1, \ldots, n$ and $j = 1, \ldots, k$. Then

$$\operatorname{tr. deg.}_C C(x_1, \ldots, x_n, y_1, \ldots, y_n) \geq n + \operatorname{rank}_K (D_i x_j)$$

unless the x_i are linearly dependent over $\mathbb{Q}$ mod C (or, equivalently, the y_i are multiplicatively dependent mod C).

Another version of the theorem is framed in [20] in terms of power series.

Ax–Schanuel is used in [45] to show that raising to a suitably generic power satisfies an analogue of Schanuel's conjecture. For a version of Ax–Schanuel in positive characteristic, see [315].

The Seidenberg Embedding Theorem

Fields of complex-analytic functions on a domain in several variables provide examples of differential fields over $\mathbb{C}$. Seidenberg's Embedding Theorem shows these indeed give all finitely generated (as differential fields) examples.

13.11 Theorem (Seidenberg [481, 482]) *Let K be a finitely generated differential field, containing $\mathbb{Q}$, with derivations $D_1, \ldots, D_m$. Then there is a domain $U \subset \mathbb{C}^m$ such that K is isomorphic as a differential field to a subfield of the field of meromorphic functions on U with derivations $\partial/\partial z_i, i = 1, \ldots, m$.*

A precise version is stated in [471, 2.6]. The differential version of Ax's theorem clearly implies the complex version, but with the above we may show the reverse implication.

13.12 Proof of Theorem 13.10 from Theorem 13.7

Suppose $K, D_1, \ldots, D_m$ with $x_i, y_i, i = 1, \ldots, n$ with $D_j y_i = y_i D_j x_i$ for all i, j and

$$\text{tr. deg.}_C C(x_i, y_i) < n + \text{rank}(D_j x_i).$$

Let W be an algebraic variety over C with $(x_1, \ldots, x_n, y_1, \ldots, y_n) \in W$. Keeping x_i, y_i and the constants required to define W, we may suppose that K is finitely generated as a differential field. By Seidenberg's embedding theorem (with the isomorphism denoted by $z \mapsto \bar{z}$) we get a domain $A \subset \mathbb{C}^m$ and meromorphic functions $\bar{x}_i, \bar{y}_i$ with $(\bar{x}, \bar{y}) \in \overline{W}$. Locally, we have $\bar{y}_i = \exp(\bar{x}_i + c_i)$ for some complex constants c_i. Now $e^{c_i} \in \mathbb{C}$ and we find that

$$\text{tr. deg.}_{\mathbb{C}} \mathbb{C}\big(\bar{x}_i, \exp(\bar{x}_i)\big) < n + \dim U,$$

where U is the locus of the $\bar{x}_i$ in $\mathbb{C}^n$, which has dimension

$$\dim U = \text{rank}\Big(\frac{\partial}{\partial z_j} \bar{x}_i\Big) = \text{rank}\big(D_j x_i\big).$$

We conclude that the $\bar{x}_i$ are linearly dependent over $\mathbb{Q}$ mod C, but the constant is clearly in the field generated by the images of the x_i, and hence the x_i are linearly dependent over $\mathbb{Q}$ mod C. □

Semi-Abelian Ax–Schanuel

Ax [21] proved a more general version of Ax–Schanuel for commutative algebraic groups and their Lie algebras in the complex setting, using differential geometric methods. This implies the analogue of complex Ax–Schanuel for semi-abelian varieties (and more generally, see [50]). See also [104] for results on elliptic curves and their extensions.

The differential version in the semi-abelian setting is deduced in [304], and applied to study the axiomatization and model-theoretic properties of the corresponding reducts of differential fields.

A Lindemann–Weierstrass Theorem for semi-abelian varieties over function fields is established in [56]; see also [50, 51].

14

A Formal Setting

We next consider various formal implications and equivalences of Ax–Schanuel related properties, for which we introduce a formal setting. Similar settings have been used (e.g. [508, 510]) to convey the analogy between Manin–Mumford and André–Oort. A more precise such setting of *distinguished categories* is provided by [40].

The Zariski-optimal form of two-sorted Ax–Schanuel (see Property 14.17) is required in Chapter 22.

Algebraic Domains and Designated Collections

14.1 Definition   By an *algebraic domain* we mean a connected semi-algebraic open subset U of a projective algebraic variety P. There is then a smallest projective subvariety of P containing U that is denoted $\widehat{U}$. If U is an algebraic domain, then an *algebraic subvariety* of U means an irreducible complex analytic component of $W \cap \widehat{U}$ for some algebraic $W \subset P$.

14.2 Definition   Let U be an algebraic domain. By a *designated collection* on U we mean a collection $\mathcal{M} = \{M\}$ of algebraic subvarieties $M \subset U$ with the following properties:

1. $U \in \mathcal{M}$.
2. $\mathcal{M}$ is closed under taking irreducible components of intersections.

14.3 Definition   Given a special collection $\mathcal{M}$ on an algebraic domain U and a subvariety $W \subset U$, condition 14.2.2 ensures there is a smallest $M \in \mathcal{M}$ with $W \subset M$, which we call the $\mathcal{M}$-*closure* of W and denote $\langle W \rangle_{\mathcal{M}}$.

Several designated collections are associated with a mixed Shimura variety X and its uniformizing map $u \colon \Omega \to X$. The collections of special and weakly special subvarieties on X, and both on Ω, are designated collections. Note also

119

that the collection of algebraic subvarieties of U is a special collection on any algebraic domain U, which we denote $\mathcal{Z}$, and that the $\mathcal{Z}$-closure is just the Zariski closure in case U is a quasi-projective variety.

Ax–Lindemann Property

We define the Ax–Lindemann property of a suitable map between algebraic domains with designated collections, and the two forms of Ax–Lindemann as in Ingredient 3.10 and Theorem 3.8 are equivalent in our general setting. The uniformizing map

$$u : \Omega \to X$$

respects the collections of weakly special subvarieties $\mathcal{L}$ on Ω and $\mathcal{T}$ on X in certain ways. If $L \in \mathcal{L}$, then $u(L) \in \mathcal{T}$, with $\dim u(L) = \dim L$, and if $T \in \mathcal{T}$, then all components U of $u^{-1}(T)$ are in $\mathcal{L}$ and have $\dim U = \dim T$.

14.4 Definition Let L_0, T_0 be algebraic domains with designated collections $\mathcal{L}, \mathcal{T}$. A *gallant map* from $\mathcal{L}$ to $\mathcal{T}$ is a surjective holomorphic map $f : L_0 \to T_0$ such that

1. If $L \in \mathcal{L}$, then $f(L) \in \mathcal{T}$.
2. If $T \in \mathcal{T}$ and $W \subset_{\mathrm{cpt}} f^{-1}(T)$, then $W \in \mathcal{L}$.
3. If $L_1 \subsetneq L_2$ is a proper subvariety, then $f(L_1) \subsetneq f(L_2)$ is a proper subvariety.

If f is a gallant map from $\mathcal{L}$ to $\mathcal{T}$, we write $f : \mathcal{L} \to \mathcal{T}$.

14.5 Proposition *Let L_0, T_0 be algebraic domains with designated collections $\mathcal{L} = \{L\}, \mathcal{T} = \{T\}$, and suppose $f : L_0 \to T_0$ is a gallant map from $\mathcal{L}$ to $\mathcal{T}$. Then the following assertions are equivalent:*

AL1. *For algebraic $W \subset L_0$, we have $\langle f(W) \rangle_{\mathcal{Z}} = f(\langle W \rangle_{\mathcal{L}})$.*
AL2. *For algebraic $V \subset T_0$, a maximal algebraic $W \subset f^{-1}(V)$ is in $\mathcal{L}$.*
AL3. *For algebraic $W \subset L_0$, we have $\langle f(W) \rangle_{\mathcal{Z}} \in \mathcal{T}$.*

Proof We show AL1 $\Rightarrow$ AL2 $\Rightarrow$ AL3 $\Rightarrow$ AL1.

Suppose that AL1 holds. Let $V \subset T_0$ and suppose $W \subset f^{-1}(V)$ is maximal algebraic. By AL1, $f(W)$ is Zariski-dense in $f(\langle W \rangle_{\mathcal{L}})$, and so $f(\langle W \rangle_{\mathcal{L}}) \subset V$, whence $W = \langle W \rangle_{\mathcal{L}}$ by maximality. Thus, AL2 holds.

Suppose that AL2 holds. Let $W \subset L_0$. Put $V = \langle f(W) \rangle_{\mathcal{Z}}$. By AL2, $W \subset L$ for some $L \in \mathcal{L}$ with $T = f(L) \subset V$. But then $\langle f(W) \rangle_{\mathcal{Z}} \subset T$, so $T = V$, and AL3 holds.

Suppose that AL3 holds. Let $W \subset L_0$ be algebraic and set $T = \langle f(W) \rangle_{\mathcal{Z}} \in \mathcal{T}$ by AL3. Then, $W \subset L$ for some $L \subset_{\mathrm{cpt}} f^{-1}(T)$ with $L \in \mathcal{L}$, and then L is the

$\mathcal{L}$-closure of W, otherwise the Zariski closure of $f(W)$ would be smaller. So AL1 holds. □

14.6 Proposition AL1 *is equivalent to the following variant:*

AL1$'$ *For algebraic* $W \subset L \in \mathcal{L}$, *if* $\langle f(W) \rangle_Z \neq f(L)$, *then* $\langle W \rangle_{\mathcal{L}} \neq L$.

Proof Let $W \subset L_0$ be an algebraic subvariety.
Assume AL1. Suppose $\langle f(W) \rangle_Z \neq T_0$. Then $f(\langle W \rangle_{\mathcal{L}}) = \langle f(W) \rangle_Z \neq T_0$.
Assume AL1$'$. Suppose $\langle f(W) \rangle_Z \neq f(\langle W \rangle_{\mathcal{L}})$. Then $\langle W \rangle_{\mathcal{L}} \neq L_0$, and $W \subset L$ for some proper $L \subset L_0$. Then $f(W) \subset f(L) = T$. Continue until $\langle f(W) \rangle_Z = T$. □

14.7 Definition If the conditions of Proposition 14.5 hold for the gallant map f of designated collections $\mathcal{L}, \mathcal{T}$, we say that $f \colon \mathcal{L} \to \mathcal{T}$ has the *Ax–Lindemann property*.

14.8 Remarks

1. The condition in Definition 14.4.3, which is a weak form of dimension preservation, enters in a weak way in AL3$\Rightarrow$AL1.
2. One can give a more general version in which algebraic subvarieties are replaced by suitable larger designated collections in the domain and codomain.

14.9 Proposition (Ax–Hermite–Lindemann) *Suppose that* $f \colon \mathcal{L} \to \mathcal{T}$ *has the Ax–Lindemann property. Then, if* $W \subset L_0$ *is algebraic with* $f(W) \subset T_0$ *also algebraic, then* $W \in \mathcal{L}$ *(and so* $f(W) \in \mathcal{T}$).

Proof This is immediate from AL3 as $f(W) = \langle f(W) \rangle_Z \in \mathcal{T}$. □

Ax–Schanuel Property

14.10 Definition (The Ax–Schanuel Property) Suppose that $f \colon \mathcal{L} \to \mathcal{T}$ is a gallant map of designated collections. For $L \in \mathcal{L}$ with $T = f(L)$, we denote by $D_L \subset L \times T$ the graph of the restriction of f to L. We say that $f \colon \mathcal{L} \to \mathcal{T}$ has the *Ax–Schanuel property* if the following holds.

Let $L \in \mathcal{L}$ with $T = f(L)$. Suppose $Y \subset L \times T$ is an algebraic subvariety and $U \subset_{\mathrm{cpt}} Y \cap D_L$ an irreducible component. Then,

$$\dim U = \dim Y - \dim L = \dim Y + \dim D_L - \dim(L \times T)$$

unless $\langle \pi_L(U) \rangle_{\mathcal{L}} \neq L$ (i.e. unless $\pi_L(U)$ is not $\mathcal{L}$-dense in L).

In the following we suppose that $f \colon \mathcal{L} \to \mathcal{T}$ is a gallant map that has the Ax–Schanuel property.

A special case of interest is the interaction under f between algebraic sub-varieties in L_0 and in T_0. This amounts to taking $Y = W \times V$ and a component $U \subset Y \cap f^{-1}(V)$.

14.11 Proposition (Two-Sorted Ax–Schanuel) *Let $L \in \mathcal{L}$ with $T = f(L)$. Suppose $W \subset L$ and $V \subset T$ are algebraic subvarieties, and let $U \subset_{\mathrm{cpt}} W \cap f^{-1}(V)$ be an irreducible component. Then,*

$$\dim U = \dim V + \dim W - \dim L$$

unless $\langle U \rangle_{\mathcal{L}} \neq L$ (i.e. U is not $\mathcal{L}$-dense in L).

Proof Let $Y = W \times V$ and $\overline{U}$ the graph of f on U. Then $\overline{U}$ is an irreducible component of $D_L \cap Y$, and the conclusion follows from the Ax–Schanuel property. □

Naturally, the Ax–Schanuel property implies Ax–Lindemann. In view of the equivalent formulations already established above in Propositions 14.6 and 14.11, it suffices to prove the following.

14.12 Proposition *Let $f \colon \mathcal{L} \to \mathcal{T}$ be a gallant map that has the Ax–Schanuel property. Let $L \in \mathcal{L}$ with $T = f(L)$. Suppose $W \subset L$ and $V \subset T$ are algebraic subvarieties with $V \neq T$ and $W \subset f^{-1}(V)$. Then, $\langle W \rangle_{\mathcal{L}} \neq L$.*

Proof We have that W is an irreducible component of $W \cap f^{-1}(V)$ and as $\dim V < \dim T$ we have

$$\dim W > \dim V + \dim W - \dim L.$$

The conclusion follows from Proposition 14.11. □

And as already seen (Proposition 14.9), this implies the *Ax–Hermite–Lindemann property*: The bi-algebraic subvarieties are precisely the ones in $\mathcal{L}, \mathcal{T}$. At the other extreme, the Ax–Schanuel property implies what one might term the Ax–Logarithms property.

14.13 Proposition *Let $f \colon \mathcal{L} \to \mathcal{T}$ be a gallant map that has the Ax–Schanuel property. Let $L \in \mathcal{L}$ with $T = f(L)$. Suppose $W \subset L$ and $V \subset T$ are algebraic subvarieties with $W \neq L$, and that $\tilde{V}$ is a component of $f^{-1}(V)$ with $U \subset W$. Then $\langle V \rangle_{\mathcal{T}} \neq T$.*

Proof We have that U is an intersection component of $W \cap f^{-1}(V)$, but $\dim U = \dim V > \dim V + \dim W - \dim L$. Hence, by Proposition 14.11, $\langle U \rangle_{\mathcal{L}} \neq L$ and so $\langle V \rangle_{\mathcal{T}} \neq T$. □

Otherwise put:

AxLogs1: *For algebraic* $V \subset T_0$. *If* $U \subset_{\text{cpt}} f^{-1}(V)$, *then* $\langle U \rangle_Z \subset_{\text{cpt}} f^{-1}(\langle V \rangle_{\mathcal{T}})$.

This has equivalent formulations dual to the ones for Ax–Lindemann.

14.14 Proposition *With assumptions as above, the following are equivalent:*

AxLogs1: *For algebraic* $V \subset T_0$, *if* $U \subset_{\text{cpt}} f^{-1}(V)$, *then* $\langle U \rangle_Z \subset_{\text{cpt}} f^{-1}(\langle V \rangle_{\mathcal{T}})$.
AxLogs2: *For algebraic* $W \subset L_0$, *a maximal algebraic* $V \subset f(W)$ *is in* $\mathcal{T}$.
AxLogs3: *For algebraic* $V \subset T_0$, *if* $U \subset_{\text{cpt}} f^{-1}(V)$, *then* $\langle U \rangle_Z \in \mathcal{L}$.

Proof We prove AxLogs1 $\Rightarrow$ AxLogs2 $\Rightarrow$ AxLogs3 $\Rightarrow$ AxLogs1.

Assume AxLogs1 and let $W \subset L_0$ be algebraic with $V \subset f(W)$ maximal algebraic. Let $U \subset_{\text{cpt}} f^{-1}(V)$ with $U \subset W$. Then $L = \langle U \rangle_Z \subset W$ for some $L \subset_{\text{cpt}} f^{-1}(\langle V \rangle_{\mathcal{T}})$. So $V \subset \langle V \rangle_{\mathcal{T}} \subset f(W)$. By maximality, $V = \langle V \rangle_{\mathcal{T}} \in \mathcal{T}$, hence AxLogs2 holds.

Assume AxLogs2 and let $V \subset T_0$ be algebraic with $U \subset_{\text{cpt}} f^{-1}(V)$. Let $W = \langle U \rangle_Z$. Then $V \subset f(W)$ so $V \subset T \subset f(W)$ for some $T \in \mathcal{T}$, by AxLogs2, and then $U \subset L \subset_{\text{cpt}} f^{-1}(T)$. So $U \subset L \subset W$, whence $W = L \in \mathcal{L}$, giving AxLogs3.

Assume AxLogs3 and let $V \subset T_0$ algebraic with $U \subset_{\text{cpt}} f^{-1}(V)$ and (by AxLogs3) $L = \langle U \rangle_Z \in \mathcal{L}$. Let $T = f(L)$, so $V \subset T$. Then $\langle V \rangle_{\mathcal{T}} = T$ otherwise U would be contained in some $L' \in \mathcal{L}$ property smaller than L. $\square$

Note that the Ax–Logarithm property (any of the conditions in Proposition 14.14) also implies the Ax–Hermite–Lindemann property.

14.15 Proposition *The Ax–Logarithms property implies the Ax–Hermite–Lindemann property.*

Proof Suppose $W \subset L_0$ is algebraic with $V = f(W) \subset T_0$ also algebraic. Then W is a component of $f^{-1}(V)$, and $W = \langle W \rangle_Z$, so by AxLogs3 we have $W = \langle W \rangle_Z \in \mathcal{L}$. $\square$

Zariski-Optimal Formulation of Two-Sorted Ax–Schanuel

For applications to Zilber–Pink problems it is convenient to reformulate the Ax–Schanuel property as done in [264] (but we adopt the modified terminology of [169]). This version has the nice feature that its formulation does not depend on the ambient $L, T = f(L)$.

14.16 Definition Let $f \colon \mathcal{L} \to \mathcal{T}$ be a gallant map. Let $L \in \mathcal{L}$ with $T = f(L)$ and fix $V \subset T$.

1. An *intersection component* of V is an irreducible component of $f^{-1}(V) \cap W$ for some algebraic subvariety $W \subset L$.
2. An intersection component A of V with Zariski closure $\langle A \rangle_Z$ has *Zariski defect* given and denoted by

$$\delta_Z(A) = \dim\langle A \rangle_Z - \dim A.$$

3. An intersection component is called *Zariski-optimal for V* if it is maximal among intersection components for its defect.
4. An intersection component A of V is called *geodesic* if A is a component of $f^{-1}(V) \cap \langle A \rangle_Z$ and $\langle A \rangle_Z \in \mathcal{L}$.

Where no ambiguity should arise, we may drop the "of/for V".

14.17 Definition (Zariski-optimal Ax–Schanuel property)   Let $f \colon \mathcal{L} \to \mathcal{T}$ be a gallant map. We say that f has the *Zariski-optimal Ax–Schanuel property* if the following holds.

A Zariski-optimal component for $V \subset T_0$ is geodesic.

Now we prove that the Zariski-optimal Ax–Schanuel property is equivalent to the Two-sorted Ax–Schanuel property.

14.18 Proof of Equivalence of Proposition 14.11 and Definition 14.17
1. Assume that the Two-Sorted Ax–Schanuel property holds (for $f, \mathcal{L}, \mathcal{T}$), and that $A \subset W \cap f^{-1}(V)$ is an optimal intersection component for V, with $W = \langle A \rangle_Z$. Let $L = \langle A \rangle_{\mathcal{L}}$ and $T = f(L)$. Let V' be the component of $T \cap V$ containing $f(A)$. Then A is Zariski-optimal for V', for if it were not, it would also fail to be Zariski-optimal for V. So we can assume $V = V' \subset T$. Since $L = \langle A \rangle_{\mathcal{L}}$, by Definition 14.10 we must have

$$\dim A = \dim W + \dim V - \dim L.$$

Now let B be the component of $f^{-1}(V)$ (in L) containing A. Then, also $\langle B \rangle_{\mathcal{L}} = L$ (as $A \subset B$) and so, by 14.10,

$$\dim B = \dim V + \dim\langle B \rangle_Z - \dim L.$$

But $\dim B = \dim V$, whence $\dim\langle B \rangle_Z = \dim L$ and so $\langle B \rangle_Z = L$, and B is a geodesic intersection component. Now

$$\delta_Z(A) = \dim W - \dim A = \dim L - \dim V = \delta_Z(B),$$

whence $A = B$ by the assumption that A is Zariski-optimal.

2. Assume that the Zariski-optimal Ax–Schanuel property holds, and let $L \in \mathcal{L}$ with $T = f(L)$. Suppose that $V \subset T, W \subset L$ are algebraic subvarieties and

that $A \subset_{\text{cpt}} W \cap f^{-1}(V)$ with $\langle A \rangle_{\mathcal{L}} = L$. There is some optimal intersection component $B \supset A$, and by Definition 14.17, it is geodesic. But $\langle B \rangle_{\mathcal{L}} = L$, and so B must be a component of $f^{-1}(V)$ with $\langle B \rangle_{\mathcal{Z}} = L$ and by 14.16 we have

$$\dim W - \dim A \geq \delta_{\mathcal{Z}}(A) \geq \delta_{\mathcal{Z}}(B) = \dim L - \dim V,$$

which re-arranges to $\dim A \leq \dim W + \dim V - \dim L$ and then equality must hold. $\qquad \square$

15

Modular Ax–Schanuel

We have already stated Modular Ax–Lindemann, Theorem 5.6, in connection with Modular André–Oort. A version with derivatives is established in [428]. Some very special cases of both were established earlier in [3].

Modular Ax–Logarithms for curves is established in [263] in connection with certain special cases of the Zilber–Pink conjecture for curves in $Y(1)^n$ (see Chapter 21). The proof uses Hodge-theoretic results of André [5] related to Deligne's Theorem of the fixed part.

15.1 Theorem ([263, Theorem 3.1]) *Let $V \subset Y(1)^n$ be a curve and let $x \in j^{-1}(V)$. Suppose that there is a proper algebraic subvariety of $\mathbb{H}^n$ that contains a neighbourhood of z in $j^{-1}(V)$. Then, V is contained in a proper weakly special subvariety of $Y(1)^n$.*

Modular Ax–Lindemann and Modular Ax–Logarithms are both special cases of Ax–Schanuel for the modular function, established in [436]. There are three versions. The basic version involves just the graph of (cartesian powers of) the modular function. A second, stronger version involves also the first and second derivatives, and is equivalent to (third version) a suitable statement in a differential field.

Complex Modular Ax–Schanuel

Let $D \subset \mathbb{H}^n \times Y(1)^n$ be the graph of $j \colon \mathbb{H}^n \to Y(1)^n$.

15.2 Theorem ([436, Theorem 1.1]) *Let $V \subset \mathbb{H}^n \times Y(1)^n$ be an algebraic variety, and let U be a complex-analytically irreducible component of $V \cap D$. Then*

$$\dim U = \dim V - n$$

unless the projection of U to $Y(1)^n$ *(or, equivalently, to* $\mathbb{H}^n$ *) is contained in a proper weakly special subvariety.*

Here $\dim U = \dim V - n = \dim V + \dim D - \dim \left(\mathbb{H}^n \times Y(1)^n \right)$ is the typical dimension, and $\dim U$ cannot be less. As the various deductions and equivalences were carried out above in a general setting, we observe that the two-sorted version and its optimal formulation (see Definition 14.17) also hold.

15.3 Theorem (Optimal form of two-sorted modular Ax–Schanuel)   *Let* $V \subset Y(1)^n$. *A Zariski-optimal intersection component for V is geodesic.*

This is the statement that is needed in general for applications to the Zilber–Pink conjecture; see Chapter 22.

Modular Ax–Schanuel with Derivatives

The modular function satisfies a third-order algebraic differential equation, and, by a result of Mahler [341], none of lower order.

15.4 Proposition (See, e.g. [350, p. 20])   *The modular function satisfies the differential equation*

$$F(j, j', j'', j''') = Sj + \frac{j^2 - 1968j + 2654208}{2j^2(j - 1728)^2} \left(j' \right)^2 = 0,$$

where ′ *denotes differentiation with respect to the argument z and Sf is the Schwarzian derivative defined as*

$$Sf = \frac{f'''}{f'} - \frac{3}{2} \left(\frac{f''}{f'} \right)^2.$$

The derivation of the differential equation from the ramification properties of j and the angles of the fundamental domain (and for more general *Schwarzian triangle groups*) is described in, for example, [502, §2.7].

The Schwarzian derivative has the property that

$$Sf = 0 \quad \text{if and only if} \quad f(z) = gz, \quad \text{for some } g \in \text{SL}_2(\mathbb{C}).$$

The full solution set of the equation is the set of "translates" of j under $\text{SL}_2(\mathbb{C})$, namely, the functions

$$j(gz), \quad \text{where } g \in \text{SL}_2(\mathbb{C}).$$

Note that there is no common domain for the full set of solutions.

Modular Ax–Schanuel with derivatives is most naturally formulated in terms of jet spaces. The following is a consequence of [436, Theorem 1.2].

15.5 Theorem ([436]) *Let $A \subset \mathbb{H}^n$ be an irreducible complex analytic variety. Then, considering all functions restricted to A,*

$$\text{tr. deg.}_\mathbb{C}\mathbb{C}\Big(z_1, j(z_1), j'(z_1), j''(z_1), \ldots z_n, j(z_n), j'(z_n), j''(z_n)\Big) \geq \dim A + 3n$$

unless A is contained in a proper weakly special subvariety.

Modular Ax–Schanuel in a Differential Field

Now we consider a differential field $(K, \{D_1, \ldots, D_m\})$ with constant field C. Note that in a differential field the elements do not have arguments, and the $'$ in the statement of the theorem is merely a symbol. The condition that "j'_i is the derivative of j_i with respect to z_i" is imposed by a differential equation in the hypothesis of the theorem.

15.6 Theorem ([436, Theorem 1.3]) *Let $z_i, j_i, j'_i, j''_i, j'''_i \in K^\times, i = 1, \ldots, n$ and suppose that*

$$D_k j_i = j'_i D_k z_i, \quad D_k j'_i = j''_i D_k z_i$$

for all i and k. Suppose further that

$$F(j_i, j'_i, j''_i, j'''_i) = 0$$

for $i = 1, \ldots, n$. Then,

$$\text{tr. deg.}_C C\Big(z_1, j_1, j'_1, j''_1, \ldots, z_n, j_n, j'_n, j''_n\Big) \geq 3n + \text{rank}\big(D_k z_i\big)_{i,k}$$

unless there are i, h with $i \neq h$ and $N \geq 1$ such that $\Phi_N(j_i, j_h) = 0$.

This theorem is proved by reducing it to the complex version with derivatives ([436, Theorem 1.2]) via the Seidenberg Embedding Theorem. The complex version is formulated in the setting of jet spaces, and proved using point-counting. It is also shown in [436] that the jet space version can be deduced directly from the differential version. For a different approach, see [82, 83].

The condition that $\Phi_N(j_i, j_h) \neq 0$ for $i \neq h$ and $N \geq 1$ gives the independence with respect to weakly special subvarieties (*geodesic independence*) as the possibility of a constant coordinate is already excluded in the hypothesis.

Unlike the exponential case, this condition *cannot* be equivalently imposed by $\text{GL}_2^+(\mathbb{Q})$-independence on the z_i (meaning that for $i \neq h$ we do not have $z_i = g z_h$ for some $g \in \text{GL}_2^+(\mathbb{Q})$), as our hypotheses do not distinguish the various solutions to the differential equation. For example, it is not excluded that the $z_i = z$ are all the same but that the $j_i = j(g_i z)$ for $g_i \in \text{SL}_2(\mathbb{C})$ such that the $g_i z$ are $GL_2^+(\mathbb{Q})$-independent. However, if one had $z_i = g_i z$ for such g_i, then one could have $j_i = j(z)$ for all i and the conclusion would fail in spite of the $GL_2^+(\mathbb{Q})$-independence of the z_i.

15.7 Remark Freitag–Scanlon [221] deduce strong minimality of the solution set of the differential equation for the j-function in a differential field, using Modular Ax–Lindemann with derivatives. See also [13, 16].

Modular Schanuel Conjecture

Like Ax–Schanuel for the exponential function, Modular Ax–Schanuel corresponds to a conjectural transcendence statement, generalizing the Modular Lindemann conjecture stated earlier (Conjecture 5.7).

15.8 Definition We say that $z_1, \ldots, z_n \in \mathbb{H}$ are *geodesically independent* if no z_i is quadratic and they are in different $GL_2^+(\mathbb{Q})$ orbits.

15.9 Conjecture *Suppose $z_1, \ldots, z_n \in \mathbb{H}$ are geodesically independent. Then,*

$$\text{tr. deg.}_{\mathbb{Q}} \mathbb{Q}\Big(z_1, j(z_1), j'(z_1), j''(z_1), \ldots, z_n, j(z_n), j'(z_n), j''(z_n)\Big) \geq 3n.$$

This conjecture follows from André's Generalized Grothendieck Period conjecture (it can be deduced from the consequences set out in [47]; there is a suitable reduction in transcendence degree if some z_i are quadratic, and if some are dependent). See also [49, Conjecture 1]. The strongest result known in this direction is due to Nesterenko (see [388]). We state it in terms of the Eisenstein series E_2, E_4, E_6.

15.10 Theorem ([388, Theorem 4.2′]) *Let $z \in \mathbb{H}$. Then,*

$$\text{tr. deg.}_{\mathbb{Q}} \mathbb{Q}\Big(e^{2\pi i z}, E_2(z), E_4(z), E_6(z)\Big) \geq 3.$$

This result implies the algebraic independence of π, e^π (and $\Gamma(\frac{1}{4})$) over $\mathbb{Q}$.

Modular Ax–Schanuel can be used to prove a generic version of Conjecture 15.9, see [208] (along similar lines to the generic version of SC in [45]).

16

Ax–Schanuel for Shimura Varieties

Let X be a (connected) Shimura variety corresponding to a semi-simple group G with uniformizing map

$$u: \Omega \to X$$

invariant under the action of $\Gamma \subset G(\mathbb{Z})$.

In this setting, Ax–Hermite–Lindemann, the characterization of bi-algebraic varieties, is established by Ullmo–Yafaev [512], using monodromy and Deligne–André [5].

16.1 Theorem ([512, Theorem 1.2]) *An irreducible subvariety $S \subset X$ is weakly special if and only if some (equivalently every) analytic component of $u^{-1}(S) \subset \Omega$ is algebraic.*

If X is of Hodge type, then a theorem due to Tretkoff [148] and Shiga–Wolfart [487], generalizing Schneider's Theorem (Theorem 5.8), asserts that $x \in \Omega$ and $u(x) \in X$ are both algebraic points if and only if x is a special point. Ullmo–Yafaev observe that this generalizes to Shimura varieties of abelian type.

16.2 Theorem ([148, 487]) *Let X be of abelian type. Suppose $x \in \Omega(\overline{\mathbb{Q}})$ and $u(x) \in X(\overline{\mathbb{Q}})$. Then x is special (also conversely).*

This leads to a corresponding characterization of special subvarieties.

16.3 Theorem ([512, Theorem 1.4]) *Let X be of abelian type. An irreducible subvariety $S \subset X$ is special if and only if it is defined over $\overline{\mathbb{Q}}$ and some (equivalently every) analytic component of $u^{-1}(S) \subset \Omega$ is algebraic over $\overline{\mathbb{Q}}$.*

We have seen that Ax–Lindemann in this setting comes up in the point-counting approach to André–Oort, and was proved in successive steps with the general case in [309]. Ax–Schanuel for Shimura varieties is established in Mok–Pila–Tsimerman [377].

16.4 Theorem ([377, Theorem 1.1]) *Let $D \subset \Omega \times X$ be the graph of the uniformizing map u. Let $V \subset \Omega \times X$ be algebraic, and let U be a complex-analytically irreducible component of $D \cap V$. Then*

$$\dim U = \dim V - \dim X$$

unless the projection of U to X (or, equivalently, to Ω) is contained in a proper weakly special subvariety.

The typical dimension $\dim V - \dim X$ is again minimal (see e.g. [337, III.4.6]). The proof of this theorem is given in §16.10, adapting the exposition in [377].

The next version includes derivatives, and a first step is to determine how many derivatives are algebraically independent. In the case of $\mathcal{A}_g$ the answer is given by Bertrand–Zudilin in [58]. Here we take z_{ij} to be the natural matrix coordinates on Siegel space $\mathbb{H}_g$, and differentials

$$\delta_{j,\ell} = \frac{1}{2\pi i}\frac{\partial}{\partial z_{j,\ell}}, 1 \le j < \ell \le g, \quad \delta_{j,j} = \frac{1}{\pi i}\frac{\partial}{\partial z_{j,j}}, 1 \le j \le g.$$

16.5 Theorem (Bertrand–Zudilin [58]) *The transcendence degree over $\mathbb{Q}$ of the differential field generated over $\mathbb{Q}$ by the field of Siegel modular functions and their partial derivatives is equal to $\dim \mathrm{Sp}_{2g} = 2g^2 + g$; the degree is the same over $\mathbb{Q}(z_{i,j})$.*

This is generalized in [377] to all Shimura varieties, though only over $\mathbb{C}$, as the (o-minimal) methods there give no control over the field of definition of the differential equations. Let $N^+ \subset G$ be the unipotent radical of an opposite parabolic subgroup of the complex parabolic subgroup $B \subset G$ defining the symmetric space. The compact dual is then given by $\widehat{\Omega} = G(\mathbb{C})/B(\mathbb{C})$.

16.6 Theorem ([377, Theorem 1.3]) *Let $z_1, \ldots, z_n$ be an $N^+(\mathbb{C})$-invariant coordinate system on Ω. Let $\phi_1, \ldots, \phi_N$ be a $\mathbb{C}$-basis of Γ-modular functions. Then the field generated by the $\{\phi_i\}$ and their partial derivatives with respect to the z_i up to order $k \ge 2$ has transcendence degree over $\mathbb{C}$ equal to $\dim G$. Further, the transcendence degree is the same over $\mathbb{C}(z_1, \ldots, z_n)$.*

The natural matrix coordinates $z_{i,j}$ on $\mathbb{H}_g$ are $N^+(\mathbb{C})$-invariant, where N^+ is the group of upper triangular matrices. Observe the implication that derivatives up to order 2 suffice to get the full transcendence degree. In fact, degree $k \ge 1$ suffices in many cases, in particular for $G = \mathrm{Sp}_{2g}, g \ge 2$; see [377, §7.3].

Ax–Schanuel with derivatives is most naturally formulated in a jet space setting, here we just state the following implication.

16.7 Theorem ([377, Theorem 1.4])   *Let $A \subset \Omega$ be an irreducible complex analytic variety. Let $\{z_i, i = 1, \ldots, n\}$ be an algebraic coordinate system on Ω. Let $\{\phi_j^{(\nu)}\}$ consist of a basis $\phi_1, \ldots, \phi_N$ of modular functions, all defined at at least one point of A, together with their partial derivatives with respect to the z_i up to order $k \geq 2$. Then, considering all functions restricted to A,*

$$\operatorname{tr.\,deg.}_{\mathbb{C}}\mathbb{C}\left(\{z_i\}, \{\phi_j^{(\nu)}\}\right) \geq \dim G + \dim A$$

unless A is contained in a proper weakly special subvariety.

There is also a version in a differential field. We refer to [377, Part III] for further details, including its equivalence with the jet space version, and here mention only the following two points.

First, in [377], the existence of algebraic differential equations (and the numerical transcendence degree in Theorem 16.6) is established via o-minimality, specifically using the Definable Chow Theorem (Theorem 12.6) and related results. This means that we do not get an explicit form for the equations, only the existence of suitable equations, including in particular the existence of a suitable *Schwarzian derivative* associated with G. For the description of a system of differential equations associated with $\mathcal{A}_g$ generalizing, see [215]. The existence of Schwarzian derivatives and algebraic differential equations associated with (suitably definable) covering maps is studied in a more general setting by Scanlon [471]. On this circle of ideas, see also [109].

Second, the condition "y satisfies the differential equation for j with respect to z" may be imposed on any non-constant elements z, y in a differential field, as the modular function is a single function of one variable. This is no longer the case for the uniformizing map on higher-dimensional Shimura varieties. Thus, these equations cannot be straightforwardly imposed on low-dimensional loci, and one must deduce the existence of suitable equations to impose on such loci. This leads to the notion of a *uniformized locus* in [377] used in formulating the differential version.

Generalizations

Mixed Ax–Schanuel

16.8 Theorem (Gao [229])   *The uniformization $\mathbb{H}_g \times \mathbb{C}^g \to \mathcal{X}_g$ of the universal family of abelian varieties (principally polarized, with suitable level structure) has the Ax–Schanuel property.*

Mixed Ax–Schanuel for the universal family of abelian varieties is useful in studying the rank of the *Betti map*, see [10, 230], which in turn plays a role in recent uniform results in diophantine geometry ([182, 231, 232]).

Ax–Schanuel for an arbitrary mixed Shimura variety follows from the more general result for variations of mixed Hodge structures referenced below. One would expect suitable analogues of the results with derivatives and differential fields to hold too. For formulations in the larger setting of *enlarged mixed Shimura varieties* (which includes universal vectorial extensions), see [228].

16.9 Variations of Hodge Structure and Other Generalizations

Bakker–Tsimerman [30] (see also [31]) generalize the Ax–Schanuel Theorem to certain variations of Hodge structure. This has found applications in different diophantine contexts; see [331, 332] (and further [184, 268] and [518, 556]). Chiu [137] and Gao–Klingler [233] generalize this and affirm Ax–Schanuel for variations of mixed Hodge structures, as conjectured in [308]. See further [138]. See [33] for a generalization to non-arithmetic ball quotients. See [82, 83] for an Ax–Schanuel result for principal bundles for an algebraic group. For Ax–Schanuel results for Lie algebras, see [402].

Proof of Theorem 16.4

16.10 Proof of Theorem 16.4

Fix a semi-algebraic fundamental domain $F \subset \Omega$ such that $u|_F : F \to X$ is definable. We assume that the image of U under projection to X is not contained in any proper weakly special subvariety and prove the dimension inequality

$$\dim U + \dim X \leq \dim V$$

(so that in fact equality holds). We adapt the exposition in [377], rearranging it to emphasize the main elements.

Volume Growth and Point-Counting As the component U is complex analytic, if it is positive-dimensional, then it has large volume. More precisely, by Hwang–To [290], the $\dim_{\mathbb{R}} U$-dimensional volume of U intersected with a ball of radius R grows exponentially with R.

On the other hand, the $\dim_{\mathbb{R}} U$-dimensional volume of

$$V \cap D \cap (\gamma \cdot F \times X)$$

is uniformly bounded in any Γ-translate $\gamma \cdot F$ of the fundamental domain F. This results from two facts. First, following [309, Lemma 5.8], using Siegel coordinates one can cover F with a finite number of sets Σ that embed into a product of (complex) one-dimensional sets of finite volume and for which the product of volumes dominates the volume in Σ. Choosing $\dim_{\mathbb{C}} U$ of these, one needs that the projection of U has finite fibres of uniformly bounded cardinality. This follows from definability.

Thus, in $B(R)$, U must go through exponentially many distinct translates $\gamma F, \gamma \in \Gamma$. By comparing distances with the natural height on Γ, it is shown in [309, §5] that the heights of the corresponding elements γ are bounded exponentially in R.

All this leads to the following conclusion.

16.11 Lemma *With the notation as above, there are positive constants c, δ such that U goes through at least*

$$cT^{\delta}$$

distinct domains $\gamma \cdot F$ with $H(\gamma) \leq T$.

We now consider the definable set

$$Z = \left\{ g \in G(\mathbb{R}) : \dim_{\mathbb{R}} \left((g \cdot V) \cap D \cap (F \times X) \right) = \dim_{\mathbb{R}} U \right\}.$$

This set contains $\gamma \in \Gamma$ whenever U goes through $\gamma \cdot F$, and we conclude from the Counting Theorem that Z contains connected positive-dimensional semi-algebraic sets with many Γ points, and in particular contains a real semi-algebraic curve C with at least two Γ points.

Let us write V_c for $c \cdot V$. Suppose V_c is non-constant as c varies. There are now two cases to consider. In the first, $U \subset V_c$ for all $c \in C$. In this case we can replace V by $V' = V_c \cap V_{c'}$ for distinct $V_c, V_{c'}$ giving an intersection with $\dim V' - \dim U$ smaller than $\dim V - \dim U$. In the second, the intersection components U_c vary. In the second we replace V by the Zariski closure V' of $C \cdot V$, which has complex dimension $\dim V' = \dim V + 1$, and intersects D in a component U' of dimension $\dim U + 1$. Both of these cases are handled by an inductive argument.

So we are left with the possibility that V_c is constant for $c \in C$, and since C contains at least two integer points we find that V is stabilized by a non-identity element of Γ. This information is difficult to work with (though it sufficed in [436]), and the next step is a step sideways to see that by moving V in a suitable family we can take it to be generic in a way which improves the specificity of the information obtained.

Varying V in an Algebraic Family Here we put V into a natural family of varieties having intersection components of the same dimension as U. We do this using the algebraicity results of Peterzil–Starchenko (see Chapter 12).

However, Mok has an approach using tools of classical complex analysis alluded to in [377, Remark p. 952].

We consider the uniformization

$$u \times \mathrm{id} \colon \Omega \times X \to X \times X,$$

where $\mathrm{id} \colon X \to X$ is the identity. It is definable on $F \times X$.

Suppose that $T \subset X \times X$ is a (relatively closed) algebraic subvariety. Then

$$\left(u \times \mathrm{id}\right)^{-1}(T) \subset \Omega \times X$$

is a closed complex-analytic set that is $(\Gamma \times \mathrm{id})$-invariant and definable on $F \times X$. The converse follows from the Definable Chow Theorem.

16.12 Proposition *Let $A \subset \Omega \times X$ be a closed complex-analytic set that is $(\Gamma \times \mathrm{id})$-invariant and definable on $F \times \Omega$. Then, $u(A) \subset X \times X$ is algebraic.*

Proof The image is closed and complex analytic in $X \times X$. It is also definable as it equals the image of $u \times \mathrm{id}$ restricted to $F \times X$. Hence, it is algebraic by the Definable Chow Theorem (Theorem 12.6). □

We now consider an algebraic variety $W_0 \subset \Omega \times X$ and a component $U \subset W_0 \cap D$ with no assumptions on dimension. We want to put W in a family of subvarieties of $\Omega \times X$ that is closed under the action of $G(\mathbb{R}) \times \mathrm{id}$ and proper. We have $\Omega \subset \widehat{\Omega}$, its compact dual, which is projective. Then W_0 is a component of

$$\widehat{W_0} \cap \left(\Omega \times X\right)$$

for some subvariety

$$\widehat{W_0} \subset \widehat{\Omega} \times \widehat{X}$$

and a natural family to work with is the Hilbert scheme M of subvarieties of $\widehat{\Omega} \times \widehat{X}$ having the same Hilbert polynomial as $\widehat{W_0}$. Then M has the structure of a projective algebraic variety, and corresponding to $y \in M$ we have the subvariety $\widehat{W_y} \subset \widehat{\Omega} \times \widehat{X}$, with one of the fibres being $\widehat{W_0}$.

We now have the incidence variety (or universal family)

$$B = \{(z, x, y) \in \Omega \times X \times M : (z, x) \in \widehat{W_y}\}$$

and the family of intersections of its fibres with D, namely,

$$A = \{(z, x, y) \in \Omega \times X \times M : (z, x) \in \widehat{W_y} \cap D\}.$$

Then A is a closed, complex analytic subset of $\Omega \times X \times M$. It has a natural projection $\theta \colon A \to M$ given by $(z, x, y) \mapsto y$ to M. For each natural number

k we consider the subset of A of points (z, x, y) such that fibre of θ containing (z, x) has dimension at least k at (z, x); that is,

$$A(k) = \{(z, x, y) \in A : \dim_{(z,x)} \theta^{-1}\theta(z, x, y) \geq k\}.$$

Then $A(k)$ is closed and complex analytic, see, for example, [412, Lemma 8.2] and references there.

Now we have the projection

$$\psi : \Omega \times X \times M \to \Omega \times X$$

and we put $Z(k) = \psi A(k)$ for each k. Then ψ is proper since M is compact and so each $Z(k)$ is a closed complex analytic subset. Also, $Z(k)$ is $(\Gamma \times \mathrm{id})$-invariant (since D is), and $Z(k) \cap (F \times X)$ is definable. Hence, by Proposition 16.12 we find the following:

16.13 Lemma   *Let $T(k) = (u \times \mathrm{id})\big(Z(k)\big) \subset X \times X$. Then $T(k)$ is a closed algebraic subvariety.*

Over any point $x \in T$ then there is some z such that, for some $y \in M_0$, (z, x, y) represents a point locally of dimension k of a subvariety in the same family as W_0. Since T contains the image of U, it too is not contained in any proper weakly special subvariety, and so has large monodromy.

Monodromy   Now we return to the V, U in Theorem 16.4, and we assume that the projection of U to Ω is not contained in any proper weakly special subvariety. We set $k = \dim U$ and construct $A(k)$ and $Z(k)$ as above, with M the Hilbert scheme of subvarieties corresponding to V.

Let $[V] \in M$ be the moduli point of $\widehat{V}$, and $A(k)' \subset A(k)$ the component containing $U \times \{[V]\}$. Let $Z(k)' = \psi(A(k)') \subset Z(k)$ be the corresponding irreducible component of $Z(k)$, and $T(k)' = (u \times \mathrm{id})(Z(k)') \subset T(k)$ the corresponding irreducible component of $T(k)$, which is then an irreducible algebraic component of $T(k)$.

By construction, $T(k)'$ contains the image of U and hence, by our assumption on U, $T(k)'$ is not contained in any proper special subvariety of the diagonal in $X \times X$. Hence, by André–Deligne [5, §5, Theorem 1] its monodromy group is dense in $G(\mathbb{R})$.

Let $M_0 \subset M$ be the family of subvarieties corresponding to $A(k)'$, and let $\Gamma_0 \subset \Gamma$ be the subgroup consisting of elements γ such that every variety in M_0 is invariant under γ. For any $\mu \in \Gamma$, denote by $E_\mu \subset M_0$ the subset corresponding to subvarieties invariant under μ. If $\mu \notin \Gamma_0$, then E_μ is a proper algebraic subvariety of M_0. Hence, a very general subvariety in the family M_0 (i.e. in the complement of a countable set of proper subvarieties, in this case, the E_μ) is invariant by precisely Γ_0.

16.14 Lemma *Let Θ be the connected component of the Zariski closure of Γ_0 in G. Then Θ is a normal subgroup of G.*

Proof There is an action of Γ on $A(k)$ given by

$$\gamma \cdot (z, x, [W]) = (\gamma \cdot z, x, [\gamma W])$$

and the map $A(k) \to Z(k)$ is equivariant with respect to this action. Since $A(k) \to Z(k)$ is proper and Γ acts discretely, the map

$$\Gamma \backslash A(k) \to \Gamma \backslash Z(k) = T(k)$$

is a proper map of analytic varieties.

Let $\Gamma \backslash A(k)'$ be the image of $A(k)'$ in $\Gamma \backslash A(k)$. Then

$$\phi \colon \Gamma \backslash A(k)' \to T(k)'$$

is a proper map of analytic varieties, and thus the fibres of ϕ have only finitely many components. Let Γ_1 be the image of

$$\pi_1(\Gamma \backslash A(k)') \to \pi_1(T(k)') \to \Gamma.$$

Now $\pi_1(\Gamma \backslash A(k)')$ has finite index in the monodromy group of $T(k)'$, so Γ_1 is still Zariski-dense in G by [5, §5, Theorem 1].

Now M_0 is invariant under Γ_1. Letting Stab(W) denote the stabilizer of W, we have that $\text{Stab}(\gamma \cdot W) = \gamma \cdot \text{Stab}(W) \cdot \gamma^{-1}$. Therefore, Γ_0 is stable under conjugation by Γ_1, and hence so also then is Θ, and then also by the Zariski closure of Γ_1, which is all of G. $\qquad\square$

Conclusion of the Proof We can now conclude the proof of Theorem 16.4. The proof is by induction, first upward on $\dim \Omega$. For a given $\dim \Omega$ we argue by induction upward on $\dim V - \dim U$, and lastly we argue by induction downward on $\dim U$.

Let us consider the base cases. First, the case $\dim \Omega = 0$ is trivial. Next consider the case that $\dim V = \dim U$. Then $V \subset D$ and the projection $u(V)$ has the same dimension as V. Then V is a component of $u^{-1}u(V)$ and hence is invariant under the monodromy group of $u(V)$. Now suppose that $u(V)$ is not contained in any proper weakly special subvariety. Then, by André–Deligne [5, §5, Theorem 1] the monodromy group of $u(V)$ is Zariski-dense in G. It follows that V is invariant under $G(\mathbb{R})^c$, which contradicts the assumption that $V \subset D$. Finally, consider the case when $\dim U = \dim X$. Then V must contain all of D, and thus V is invariant under $G(\mathbb{Z})$. Since V is algebraic, it must then be invariant under $G(\mathbb{R})^c$, and so $V = \Omega \times X$. But then the theorem holds ($\dim U$ is not excessive).

So now we consider a non-base case. By forming the family as above, we may assume that V is a very general member of the family M_0 and is invariant by precisely $\Gamma_0 \subset \Gamma$. Hence, the connected component Θ of the Zariski closure of Γ_0 in G is a normal subgroup of G, by Lemma 16.14.

We can assume that U has positive dimension, as a point is weakly special (and proper as $\dim X > 0$). Then by volume estimates and point-counting we get (as above) a real semi-algebraic $C \subset Z$, and we write V_c for $c \cdot V$ as before.

Suppose that V_c remains constant as c varies. If $U \subset V_c$ as c varies, then we can replace V by $V' = V_c \cap V_{c'}$ for distinct $c, c' \in C$ and thereby decrease $\dim V - \dim U$. Then we conclude by induction. If not, we replace V by the Zariski closure of $C \cdot V$, which intersects D in a component of dimension $\dim U + 1$. We again conclude by induction.

Finally, we have the case where V_c is constant, so that V has a positive-dimensional stabilizer Θ that, by construction, is a $\mathbb{Q}$-group and a normal subgroup of G. It follows that G is isogenous to a product $\Theta \times \Theta'$ and we have a splitting $\Omega = \Omega_\Theta \times \Omega_{\Theta'}$ of Hermitian symmetric domains. After possibly replacing Γ by a subgroup of finite index we have $X = X_\Theta \times X_{\Theta'}$.

Since V is invariant under Θ, it has the form $\Omega_\Theta \times V_1$ for some $V_1 \subset \Omega_{\Theta'} \times X_{\Theta'} \times X_\Theta$. Also D decomposes as $D_\Theta \times D_{\Theta'}$. Let U_1 be the projection of U to V_1. Since the map $D_\Theta \to X_\Theta$ has discrete fibres, it follows that $\dim U_1 = \dim U$.

Now let V' be the image of V_1 under projection to $\Omega_{\Theta'} \times X_{\Theta'}$, and U' the component of $V' \cap D_{\Theta'}$ containing (and hence equal to) the projection of U to $\Omega_{\Theta'} \times X_{\Theta'}$.

Then the projection of U' to $X_{\Theta'}$ is not contained in any proper weakly special subvariety of $X_{\Theta'}$. Let V'' be the Zariski closure of U'. By induction we have

$$\dim U' + \dim X_{\Theta'} \le \dim V''.$$

Consider the projection $\psi : W_1 \to W'$. The generic fibre dimension over $W'' \subset W'$ is the same as it is over U', so

$$\dim U_1 + \dim X_{\Theta'} \le \dim \psi^{-1}(V'') \le \dim V_1$$

and it follows from this that

$$\dim U + \dim X \le \dim V.$$

This completes the induction and the proof of Theorem 16.4. $\square$

Related Results

While Ax–Lindemann describes the Zariski closure of the image $u(W) \subset X$ of an algebraic subvariety $W \subset \Omega$, the topological closure $u(W)^{\text{top}}$ is considered in [515] in the case of the uniformization $u \colon \mathbb{C}^g \to \Lambda \backslash \mathbb{C}^g = A$ of an abelian variety. It is proved that when W is a curve, $u(W)^{\text{top}}$ is the union of $u(W)$ with a finite number of *real weakly special subvarieties*. These are real analytic subvarieties which are (up to translation) the image of a real vector subspace of $\mathbb{C}^g$ that intersects the lattice Λ in a lattice. See also [183].

The Zariski closure of the image in A of a definable subset $Z \subset \mathbb{C}^g$ is studied in [516]. Under suitable additional conditions, they prove that the image contains a Zariski-dense set of positive-dimensional weakly special subvarieties.

A full description of the topological closure in general in both cases is given in [417], using more tools from the model theory of valued fields and o-minimality. See further [418]. It is an interesting question to establish suitable analogues of these results in Shimura varieties.

Using point-counting in a way similar to the proof of Ax–Lindemann, an analogue of the Bloch–Ochiai Theorem for Shimura varieties is established in [242] (the case of a compact Shimura variety was established earlier in [517]). For a value distribution approach to such problems, see [406].

17

Quasi-Periods of Elliptic Curves

The objective of this chapter is to show that Theorem 15.6 (Modular Ax–Schanuel in a differential field) may be reinterpreted as an Ax-type version of the elliptic curve case of André's Generalized Period conjecture (see §13.6).

Setting

The Grothendieck Period conjecture (PC) and André's generalized version in the case of 1-motives are discussed in detail by Bertolin [47]. We state a special case. Let E be an elliptic curve in Weierstrass form

$$E: y^2 = 4x^3 - g_2 x - g_3,$$

where $g_2, g_3 \in \mathbb{C}$, with discriminant $\Delta = g_2^3 - 27g_3^2 \neq 0$. Its j-invariant is $j = 12^3 g_2^3 / \Delta$. If $g_2, g_3 \in \overline{\mathbb{Q}}$ and E does not have CM, then PC predicts that

$$\text{tr. deg.}_{\mathbb{Q}} \mathbb{Q}(\omega_1, \omega_2, \eta_1, \eta_2) \geq 4,$$

where ω_1, ω_2 are two linearly independent periods of the regular differential (see formulary below), and η_1, η_2 are corresponding quasi-periods. (In the CM case, the expected transcendence degree reduces to 2 and the result is a theorem of Chudnovsky; see [47].)

Moreover, if $E^{(1)}, \ldots, E^{(n)}$ are elliptic curves, then PC predicts that

$$\text{tr. deg.}_{\mathbb{Q}} \mathbb{Q}(\omega_1^{(i)}, \omega_2^{(i)}, \eta_1^{(i)}, \eta_2^{(i)}) \geq 3n + 1$$

unless there exist i, k such that $E^{(i)}$ and $E^{(k)}$ are isogenous; that is, $\Phi_N(j^{(i)}, j^{(k)}) = 0$ for one of the classical modular polynomials $\Phi_N(X, Y)$, or some $E^{(i)}$ has CM (in these cases there is a corresponding reduction in the predicted transcendence degree: see [47]). Note that due to the Legendre relation $\omega_2 \eta_1 - \omega_1 \eta_2 = 2\pi i$, this is the maximum possible transcendence degree.

André's generalized version predicts, in this setting, that, for any $g_2^{(i)}$, $g_3^{(i)} \in \mathbb{C}$,

$$\mathrm{tr.deg.}_{\mathbb{Q}}\mathbb{Q}\big(g_2^{(i)}, g_3^{(i)}, \omega_1^{(i)}, \omega_2^{(i)}, \eta_1^{(i)}, \eta_2^{(i)}\big) \geq 3n + 1 \qquad (+)$$

unless some $E^{(i)}, E^{(k)}$ with $i \neq k$ are isogenous or some $E^{(i)}$ has CM.

Our objective is to establish a differential analogue of the above statement (+). It turns out that such a statement is equivalent to Modular Ax–Schanuel with derivatives.

Formulary

We begin with some formulary following [135]. We set $J = J(E) = j/12^3$. The curve E as above is isomorphic to the curve

$$E_J : y^2 = 4x^3 + g(x + 1), \qquad g = -\frac{g_2^3}{g_3^2}.$$

The periods Ω_1, Ω_2 of E_J may be defined using an explicit choice of loops in a certain region of the parameters (see [135]). They satisfy the following Picard–Fuchs differential equation:

$$\frac{d^2\Omega}{dJ^2} + \frac{1}{J}\frac{d\Omega}{dJ} + \frac{31J - 4}{144J^2(1 - J)}\Omega = 0. \qquad (*)$$

The corresponding *quasi-periods* H_1, H_2 of E_J are given by

$$H = \frac{3}{2}\frac{J + 2}{J - 1}\Omega - 18J\frac{d\Omega}{dJ} \qquad (**)$$

and we further have (see [245, Equation 4.8])

$$\Omega = \frac{2(2 + J)}{3J}H - 8(1 - J)\frac{dH}{dJ} :$$

these last two relations give (*). For E itself we then have

$$\omega_1 = \sqrt{\frac{g_2}{g_3}}\Omega_1, \quad \omega_2 = \sqrt{\frac{g_2}{g_3}}\Omega_2, \quad \eta_1 = \sqrt{\frac{g_3}{g_2}}H_1, \quad \eta_2 = \sqrt{\frac{g_3}{g_2}}H_2.$$

This then defines all the quantities appearing above.

Differential Setting and Main Result

We now consider a differential field $(K, \{D_1, \ldots, D_m\}, C)$, where K is a field of characteristic zero, D_i are a finite set of commuting derivations on K, and C is the constant field. By the *rank* of some elements $y_1, \ldots, y_m \in K$ (over K) we mean the K-rank of the matrix $(D_i y_k)$. For a subfield $L \subset K$ we define

$\mathrm{rk}(L) = \mathrm{rk}_K(L)$ to be the maximal rank of any subset, or (equivalently) the rank of a set of generators of L as a field.

Let $x, y \in K$. We say that y *is a function of* x if they are both constant or x is non-constant and Dy/Dx is independent of $D \in \mathcal{D}$. The common value is then denoted dy/dx, and one can check that then dy/dx is also a function of x.

By an *elliptic curve* over K we mean an elliptic curve E/K in Weierstrass form

$$E : y^2 = 4x^3 - g_2 x - g_3, \quad g_2, g_3 \in K, \quad \Delta = g_2^3 - 27g_3^2 \neq 0, \quad j = 12^3 g_2^3/\Delta.$$

The case in which j is constant for one of our curves is exceptional, and we may therefore assume that j is non-constant (and say that E is *non-constant*). Thus, the CM case does not arise for us. Then at least one of g_2 and g_3 must also be non-constant, and both must be non-zero, so we may define E_J as above.

By a *period* of E_J we mean some $\Omega \in K$ that is a function of J (which is also non-constant) and satisfies the differential equation

$$\frac{d^2\Omega}{dJ^2} + \frac{1}{J}\frac{d\Omega}{dJ} + \frac{31J - 4}{144J^2(1 - J)}\Omega = 0. \qquad (*)$$

This equation, considered in meromorphic functions, has locally two linearly independent solutions, so we may assume that two such solutions belong to K.

We then define the corresponding quasi-period using (**) as above. The solutions to (*) in K form a two-dimensional vector space over C. If two solutions are linearly dependent, then their ratio lies in C. Thus, linear independent of the solutions amounts to linear dependence over C and independence can be imposed using the Wronskian determinant.

Here is a first version of the differential analogue of (+).

17.1 Theorem   *Let $E^{(1)}, \ldots, E^{(n)}$ be non-constant elliptic curves over K.*

For each i, let $\Omega_1^{(i)}, \Omega_2^{(i)} \in K$ be two solutions of equation (*) that are linearly independent over C. Define $H_1, H_2 \in K$ using (**) and set

$$\omega_1 = \sqrt{\frac{g_2}{g_3}}\Omega_1, \quad \omega_2 = \sqrt{\frac{g_2}{g_3}}\Omega_2, \quad \eta_1 = \sqrt{\frac{g_3}{g_2}}H_1, \quad \eta_2 = \sqrt{\frac{g_3}{g_2}}H_2.$$

Let

$$M = C\left(g_2^{(i)}, g_3^{(i)}, \omega_1^{(i)}, \omega_2^{(i)}, \eta_1^{(i)}, \eta_2^{(i)}\right).$$

Then

$$\mathrm{tr.\,deg.}_C(M) \geq 3n + \mathrm{rk}_K(M)$$

unless $\Phi_N(j^{(i)}, j^{(k)}) = 0$ for some $N \geq 1$ and $i \neq k$.

Note that since $\omega_2\eta_1 - \omega_1\eta_2$ is a constant, this is again the maximum possible general lower bound for the transcendence degree since one could have each

$g_2^{(i)}, g_3^{(i)}$ algebraic over $j^{(i)}$ and the total transcendence degree of the $j^{(i)}$ could be as small as the rank.

If we assume additionally that $C \subset \mathbb{C}$, then we can take $\pi \in C$. Then, among the functions generating M/C we have at least $3n + \mathrm{rk}_K(M)$ that are algebraically independent over C, and hence over $\mathbb{Q}(\pi)$. As tr. deg.$_\mathbb{Q}(\mathbb{Q}(\pi)) = 1$ we conclude the following.

17.2 Corollary *Suppose, in addition to the hypotheses of Theorem 17.1, that we have $\mathbb{Q}(\pi) \subset C \subset \mathbb{C}$. Then,*

$$\text{tr. deg.}_\mathbb{Q} \mathbb{Q}\left(\pi, g_2^{(i)}, g_3^{(i)}, \omega_1^{(i)}, \omega_2^{(i)}, \eta_1^{(i)}, \eta_2^{(i)}\right) \geq 3n + 1 + \mathrm{rk}_K(M),$$

unless $\Phi_N(j^{(i)}, j^{(k)}) = 0$ for some $N \geq 1$ and $i \neq k$.

In the setting of Corollary 17.2, we can further normalize the periods and quasiperiods to satisfy the Legendre relation and recover a full functional analogue of (+).

17.3 Corollary *Suppose, in addition to the hypotheses of Corollary 17.2, that we have $\omega_2^{(k)}\eta_1^{(k)} - \omega_1^{(k)}\eta_2^{(k)} = 2\pi i$ for each k. Then,*

$$\text{tr. deg.}_\mathbb{Q} \mathbb{Q}\left(g_2^{(i)}, g_3^{(i)}, \omega_1^{(i)}, \omega_2^{(i)}, \eta_1^{(i)}, \eta_2^{(i)}\right) \geq 3n + 1 + \mathrm{rk}_K(M),$$

unless $\Phi_N(j^{(i)}, j^{(k)}) = 0$ for some $N \geq 1$ and $i \neq k$.

Proofs

We deduce Theorem 17.1 from a reformulation of Theorem 15.6. Instead of considering j as a function of z we may consider z as a function of j. The derivatives $(dj/dz, d^2j/dz^2, \ldots, dz/dj, d^2z/dj^2, \ldots)$ are algebraic over each other (easy elementary check) and we obtain the following version equivalent to Theorem 15.6.

17.4 Theorem *Let*

$$j^{(i)}, z^{(i)}, z^{(i)'}, z^{(i)''}, z^{(i)'''} \in K, \quad i = 1, \ldots, n,$$

where, for each $i = 1, \ldots, n$, we have that $j^{(i)}$ is non-constant, $z^{(i)}$ is a function of $j^{(i)}$, and $z^{(i)'}, z^{(i)''}, z^{(i)'''}$ denote differentiation with respect to $j^{(i)}$. Suppose

$$S_{j^{(i)}} z^{(i)} + R(j^{(i)}) = 0, \quad R(x) = \frac{x^2 - 1968x + 2654208}{2x^2(x - 1728)^2}$$

for each i. Let

$$L = C\left(j^{(i)}, z^{(i)}, z^{(i)'}, z^{(i)''}, z^{(i)'''}\right).$$

Then,

$$\text{tr. deg.}_{\mathbb{C}}(L) \geq 3n + \text{rk}_K(L)$$

unless there exists $i \neq k$ such that $\Phi_N(j^{(i)}, j^{(k)}) = 0$ for some $N \geq 1$.

Proof of Theorem 17.1 Suppose we are given a field M as in the said theorem. Then, for each i, and temporarily suppressing the superscript "(i)", the ratio $z = \omega_1/\omega_2 = \Omega_1/\Omega_2$ satisfies the j-differential equation with respect to j. The derivatives z', z'', z''' also are algebraic over M. Thus, if we let

$$L = \mathbb{C}\left(j^{(i)}, z^{(i)}, z^{(i)'}, z^{(i)''}, z^{(i)'''}\right),$$

we have, by Theorem 17.4, that

$$\text{tr. deg.}_{\mathbb{C}}(M) \geq \text{tr. deg.}_{\mathbb{C}}(L) \geq 3n + \text{rk}_K(L)$$

unless the exceptional case in Theorem 17.4 holds, which is the same as the exceptional case in Theorem 17.1.

Now if $\text{rk}_K(M) = \text{rk}_K(L)$, we have the conclusion of Theorem 17.1. However, if $\text{rk}_K(M) > \text{rk}_K(L)$, then the difference in transcendence degrees is just as great, namely,

$$\text{tr. deg.}_{\mathbb{C}}(M) - \text{tr. deg.}_{\mathbb{C}}(L) \geq \text{rk}_K(M) - \text{rk}_K(L).$$

That completes the proof. □

Conversely, we can deduce Theorem 17.4 from Theorem 17.1 so that Modular Ax–Schanuel with derivatives is equivalent to this Ax-Generalized-Period-conjecture for products of elliptic curves.

Proof of Theorem 17.4 from Theorem 17.1 Now we suppose that Theorem 17.1 holds, and that

$$L = \mathbb{C}\left(j^{(i)}, z^{(i)}, z^{(i)'}, z^{(i)''}, z^{(i)'''}\right)$$

is a field as in Theorem 17.4. We may choose $g_2^{(i)}, g_3^{(i)}$ algebraic over $j^{(i)}$ so that $j^{(i)}$ is their j-invariant.

For each i we seek (dropping the superscript "(i)") $\Omega_1, \Omega_2, H_1, H_2$ as in Theorem 17.1 for which we have $z = \Omega_1/\Omega_2$, etc. A computation finds that

$$18Jz' = \frac{\Omega_1 H_1 - \Omega_1 H_2}{\Omega_2^2} = \frac{c}{\Omega_2^2}$$

for some constant c. So we can choose some c (which is undetermined by our requirements), then Ω_2 is algebraic over z', J, then $\Omega_1 = z\Omega_2$ and finally $(18z'J)'$ computes H_2 in terms of the other quantities. □

PART IV

THE ZILBER–PINK CONJECTURE

18

Sources

The Zilber–Pink conjecture (ZP) is an open diophantine conjecture that unifies the Mordell–Lang conjecture and the André–Oort conjecture, but also far extends them. It was first proposed (in the semi-abelian setting) by Zilber [550], then independently by Bombieri–Masser–Zannier [90], and finally formulated in the general setting of mixed Shimura varieties by Pink [445]. There are slight differences in the formulations (discussed further below), and we take "the" Zilber–Pink conjecture to be the strongest form: the conjecture proposed by Zilber and Bombieri–Masser–Zannier (which are equivalent) in the settings considered by Pink.

In this chapter, we trace the evolution of ZP from the Mordell conjecture.

Mordell, Manin–Mumford, and Mordell–Lang Conjectures

18.1 Conjecture (Mordell conjecture [212, 381]) *A smooth irreducible curve of genus at least 2 defined over a number-field K has only finitely many K-rational points.*

The first step in the development, due to Lang in the period when the conjecture was still open, recasts and extends the conjecture in terms of a subvariety of a group variety. Mordell proved (in the same paper, [381]) that the group of rational points on an elliptic curve is finitely generated. This was generalized to abelian varieties and arbitrary number-fields by Weil. A curve possessing a rational point over a number-field K may be embedded over K into its Jacobian A, an abelian variety, in which $A(K)$ is a finitely generated group, and now one can generalize the Mordell conjecture as follows ([328, I §6, p. 36]):

Let $V \subset A$ be a curve of genus at least 2 embedded in an abelian variety and $\Gamma \subset A$ a finitely generated subgroup. Then $V \cap \Gamma$ is finite.

Similar formulations had already arisen in work on the unit equation in work of Siegel, Mahler, Davenport–Mahler; see, for example, [327, Notes and Comments to §8]. In this guise the statement is similar to the Manin–Mumford conjecture that, in its original form, asserted finiteness for the torsion points on V in such an embedding. Lang combined these into the Mordell–Lang conjecture, in which a subvariety is intersected with the division group of a finitely generated subgroup.

18.2 Definition   Let A be an abelian variety or a power of $\mathbb{G}_m$ (or more generally a semi-abelian variety) and let $\Gamma_0 \subset A$ be a finitely generated group with *division group*

$$\Gamma = \{x \in X : \exists n \in \mathbb{Z}\backslash\{0\} : n.x \in \Gamma_0\}.$$

A Γ-*special subvariety of A* is a coset of an abelian subvariety (or more generally of a connected algebraic subgroup) by a point of Γ.

Mordell–Lang comes in several versions, depending on whether one restricts the ambient variety A to be an abelian variety or a semi-abelian variety (which then includes also $\mathbb{G}_m^n$), on whether one intersects the subvariety $V \subset A$ with points defined over a number-field, or a finitely generated field, or points in a finitely generated group, or, in the strongest version, the division group of a finitely generated group. See, for example, [324], also [328, I §6.1] and the surrounding discussion, the discussion in [327, Notes and Comments to §8], and [392].

18.3 Conjecture (Mordell–Lang (ML) Conjecture [328, Conjecture I §6.3])
With A, Γ as in Definition 18.2, let $V \subset X$. Then V contains only finitely many maximal Γ-special subvarieties. We refer to the case in which $A = \mathbb{G}_m^n$ as the Multiplicative Mordell–Lang conjecture (MML; Theorem of Laurent [329]).

The Mordell–Lang conjecture is a theorem, proved in the work of a number of people, including Hindry, Laurent, Liardet, McQuillan [369], Raynaud, and Vojta, combined with Faltings Big Theorem [213, 214] (intersecting a subvariety of an abelian variety with points defined over a number-field).

André–Oort and Generalized André–Oort Conjectures

The next step was the formulation, by André and Oort, partly in analogy with the Manin–Mumford conjecture, of the André–Oort conjecture (Conjecture 6.4). Here the ambient variety is a Shimura variety equipped with its collection of special subvarieties, rather than a group variety.

The broader class of *mixed Shimura varieties* (see [226, 445]) has already been mentioned at several points. The paradigm example is the universal family X_g of (principally polarized) abelian varieties of given dimension (with suitable level structure). The simplest mixed Shimura variety of this kind is the Legendre family of elliptic curves.

18.4 Definition   The *Legendre family* $\mathcal{L} \to Y(2)$ is the family of elliptic curves depending on the parameter $\lambda \in \mathbb{P}^1 \setminus \{0, 1, \infty\}$ defined (conventionally in affine coordinates) by

$$\mathcal{L} : y^2 = x(x-1)(x-\lambda), \quad \lambda \in Y(2) = \mathbb{P}^1 \setminus \{0, 1, \infty\}.$$

Though defined in affine coordinates, the Legendre family is to be considered as a family of projective elliptic curves $E_\lambda \subset \mathbb{P}^2$, hence given by the homogenization of the above equation,

$$\mathcal{L} : ZY^2 = X(X-Z)(X-\lambda Z) \subset Y(2) \times \mathbb{P}^2,$$

as a quasi-projective surface. The base $Y(2)$ is the modular curve parameterizing elliptic curves with full level-2 structure (i.e. choice of a basis of the 2-torsion points).

18.5 Definition   The *special subvarieties* of $\mathcal{L}$ are as follows. The special points of $\mathcal{L}$ are the torsion points in CM fibres. The special subvarieties of $\mathcal{L}$, apart from $\mathcal{L}$ itself and the special points, are CM fibres E_λ, and torsion sections.

One constructs in an analogous way the universal family of (principally polarized) abelian varieties of given dimension (see, e.g. [384, Ch. II, 4.4]). As in the elliptic curve case, to get a universal family one has to take some level structure by going to a suitable finite cover of $\mathcal{A}_g$. (This is set out, in, for example, [227].) The various conjectures are independent of the choice of level structure, and we suppress mention of it in the notation.

18.6 Definition   The *universal family of (principally polarized) abelian varieties of dimension g* (with suitable level structure) is denoted X_g. One has the projection $\pi : X_g \to \mathcal{A}_g$, and the fibre of π over $x \in \mathcal{A}_g$ is denoted A_x, an abelian variety of dimension g whose moduli point is x. The *special points* on X_g are the torsion points on CM fibres.

More generally, a *mixed Shimura variety* is constructed from group-theoretic data beginning with a connected linear algebraic group over $\mathbb{Q}$; for a definition, see [444] or [226]. Mixed Shimura varieties are quasi-projective varieties defined over $\overline{\mathbb{Q}}$.

18.7 Definition ([444] or [226])   A mixed Shimura variety X has a collection $S = S_X$ of *special subvarieties*, see, for example, [444, Definition 4.1(a)] and a larger collection $\mathcal{W} = \mathcal{W}_X$ of *weakly special subvarieties*, see, for example, [444, Definition 4.1(b)]. The special subvarieties of dimension zero are called *special points*. A weakly special subvariety that contains a special point is special ([444, Proposition 4.15], and a special subvariety contains a Zariski-dense set of special points ([444, Proposition 4.14]).

The Generalized André–Oort conjecture for mixed Shimura varieties, which is just the analogue of AO, was formulated by André [7, Lecture 3], where it is proved in the basic case of $\mathcal{L}$.

18.8 Conjecture (Generalized André–Oort conjecture)   *Let X be a mixed Shimura variety, and $V \subset X$ a subvariety. Then, V contains only finitely many maximal special subvarieties.*

18.9 Theorem ([7, Lecture 3])   *Let $V \subset \mathcal{L}$ be a curve. If V is not special, then it contains only finitely many special points.*

André–Pink–Zannier Conjecture

Abelian varieties (and, as we see later, $\mathbb{G}_m$, and more generally semi-abelian varieties) appear as (in general) weakly special subvarieties of mixed Shimura varieties. Thus, mixed Shimura varieties provide a setting in which all the diophantine problems considered can be comprehended. The Generalized André–Oort conjecture implies MM for CM abelian varieties A_x as we may identify A_x with the special subvariety $\{x\} \times A_x \subset \mathcal{X}_g$ and the torsion points on A_x correspond precisely to special points on $\{x\} \times A_x$. However, the torsion points on non-CM abelian varieties are not special, and so Generalized AO for $\mathcal{X}_g$ does not imply the Manin–Mumford conjecture for a non-CM abelian variety.

A different conjecture of André [4, Ch. X, Problem 3], extended by Pink [445, Conjecture 1.6] (and similar to an unpublished conjecture of Zannier; see [227]), combines the full Mordell–Lang conjecture with a conjecture about Zariski density of points from Hecke orbits (or a slightly different version: in isogeny orbits). This is known as the André–Pink–Zannier conjecture, but it does not include the full André–Oort conjecture.

18.10 Definition   For $Y(1)^n$:

1. The *Hecke orbit* of $x \in Y(1)$ is

$$\Sigma(x) = \{y \in Y(1) : \exists N \text{ with } \Phi_N(x, y) = 0\} = j\left(\mathrm{GL}_2^+(\mathbb{Q})j^{-1}(x)\right).$$

2. The *Hecke orbit* of $x = (x_1, \ldots, x_n) \in Y(1)^n$ is

$$\Sigma(x) = \{y = (y_1, \ldots, y_n) \in Y(1)^n : y_i \in \Sigma(x_i), i = 1, \ldots, n\}.$$

3. A $\Sigma(x)$-*special subvariety* of $Y(1)^n$ is a weakly special subvariety containing a point (and hence a dense set of points) $y \in \Sigma(x)$.

18.11 Conjecture (André–Pink–Zannier; [444, Conjecture 1.6]) *Let X be a connected mixed Shimura variety ([444, 2.1]) and Σ the generalized Hecke orbit of a point $x \in X$ (see [444, 3.3]). Let $V \subset X$. If $\Sigma \cap V$ is Zariski-dense in V, then V is weakly special.*

This conjecture comes in a few versions, according to whether one considers Hecke orbits, generalized Hecke orbits (and how these are defined: see [444, Remark 3.5]; see also [459, 555]), or isogeny orbits. In the case of $\mathcal{A}_g$, an isogeny may or may not respect the polarizations. This leads to a notion of *isogeny class* of a point $x \in \mathcal{A}_g$ that is in general larger than its Hecke orbit, in which polarizations are respected. All these notions coincide with the above in $Y(1)^n$, except that the generalized Hecke orbit is closed under permutations of the coordinates.

Conjecture 18.11 implies the Mordell–Lang conjecture ([444, Theorem 5.4]; and for subvarieties of an abelian variety it is implied by ML). Thus, one might think of Conjecture 18.11 as a Generalized Mordell–Lang conjecture. But as special points do not lie in finitely many generalized Hecke orbits, it does not imply AO. Conjecture 18.11 is in turn a consequence of Pink's conjecture (Conjecture 19.19; [445, Theorem 3.3]), which follows from ZP.

Yet another prototype of unlikely intersection problems is André's Theorem [4, Theorem I (proved in Ch. X)] on endomorphisms in the fibres of an abelian pencil; see discussion of related results before Proposition 20.30).

Atypical Intersections

The final step in the evolution of ZP was taken, with different motivations, and arriving at different formulations, by Zilber, Bombieri–Masser–Zannier, and Pink. A special case had been formulated earlier in a different form by Zhang [547, §4, Remark 4a,b] (which includes a "Bogomolov" aspect); see the discussion in [89]. See also another precursor in [557, Conjecture 1] and [453, Question].

Let A, B be irreducible subvarieties of a variety X. By considering codimension as counting the number of independent conditions, one expects that components of $A \cap B$ have dimension

$$\dim A + \dim B - \dim X,$$

in particular one expects the intersection to be empty if this quantity is negative. In fact, except possibly at singular points of X, the components of $A \cap B$ always have *at least* this dimension.

18.12 Theorem ([383, 3.28, p. 57])   *Let X be an affine variety of dimension n, and let $A, B \subset X$ be subvarieties of dimension r, s, and assume x is a smooth point on X. Then if*

$$A \cap B = W_1 \cup \cdots \cup W_k \cup V_1 \cup \cdots \cup V_\ell,$$

where $x \in W_i, i = 1, \ldots, k$, but the components $V_j, j = 1, \ldots, \ell$ are not through x, then

$$\dim W_i \geq r + s - n.$$

In particular, if X is smooth, every component of $A \cap B$ has at least this dimension.

The conclusion can fail when the ambient variety is not smooth [383, p. 58]. The hypotheses evidently apply to subvarieties of $\mathbb{G}_m^n$, abelian varieties, and $Y(1)^n$, and abelian varieties, which are smooth (and Shimura varieties can be assumed to be smooth after possibly passing to a finite cover).

The crucial new element then is to consider not just special points (or subvarieties) contained in V but more generally intersections of V with a special subvariety that are atypically large in dimension.

Here we follow Pink and show how this idea connects the Manin–Mumford conjecture for a general abelian variety with unlikely intersections (when the expected dimension is negative) in the universal family of abelian varieties.

Consider the universal family $\mathcal{X}_g \to \mathcal{A}_g$ of (principally polarized) abelian varieties and a point $x \in \mathcal{A}_g$. Let $A = A_x \subset \mathcal{X}_g$ be the corresponding abelian variety, and $V \subset A$ a proper subvariety. Unless x is a special point, a torsion point P of A is not a special point of $\mathcal{X}_g$. However, it is a point of intersection with V with a torsion section $S \subset \mathcal{X}_g$, whose dimension is $\dim \mathcal{A}_g$. Now

$$\dim V < \dim A, \quad \dim A + \dim S = \dim \mathcal{X}_g$$

and so

$$0 = \dim\{P\} > \dim V + \dim S - \dim \mathcal{X}_g,$$

whence P is an unlikely intersection of V with the special subvariety S of $\mathcal{X}_g$.

The Uniform Schanuel Conjecture and Atypical Intersections

Here we describe the context in which Zilber formulated his conjecture on Intersections with Tori ([550]). This concerns the model theory of complex exponentiation.

As we have seen in Example 7.13.4, the first-order theory of the complex exponential field $\mathbb{C}_{\exp} = (\mathbb{C}, +, \times, 0, 1, \exp)$ is wild because the integers are definable in it. However, it is shown in [549, 551] that the theory has excellent properties if one works in a stronger *infinitary logic* (known as $L_{\omega_1, \omega}$, which admits certain (countably) infinitely long formulae and thereby enables the kernel of exponentiation to be fixed as a standard copy of the integers), and assumes two conjectures. The first is Schanuel's conjecture, which one sees as asserting that certain systems of algebraic-exponential equations have no solutions, and a complementary Strong Existential Closedness conjecture (SEC), formulated in [551], which asserts, roughly, that all systems have a solution unless this is precluded by SC. This is sometimes referred to as a dual SC. Moreover, SEC posits that admissible systems have solutions that are generic over any given finitely generated field. Like SC, only very special cases are known ([105, 160, 343, 346]).

A suitable $L_{\omega_1, \omega}$-axiomatization of an algebraicially closed field with an exponential map (pseudo-exponentiation) satisfying these conjectures is categorical in uncountable powers ([551]), and hence is as nice (in this sense) as the theory of algebraically closed fields. This is based on Shelah's theory of *quasi-minimal excellence*. Let $\mathbb{B}$ (known as the *Zilber field*) be a model of this theory of continuum size. Zilber conjectures that $\mathbb{B}$ and $\mathbb{C}_{\exp}$ are isomorphic, which would follow if both SC and SEC hold in $\mathbb{C}_{\exp}$. This would further imply the quasi-minimality of $\mathbb{C}_{\exp}$ as $\mathbb{B}$ is quasi-minimal.

A number of results show that properties known for $\mathbb{C}_{\exp}$ hold in $\mathbb{B}$, or vice versa. For example, closely related to SEC is the Schanuel Nullstellensatz. Let us denote by T the smallest ring of entire functions containing all complex polynomials in any number of variables and closed under composition and exponentiation. Henson–Rubel [274] prove the conjecture, due to Schanuel, that if $f \in T$ never vanishes, then $f = \exp(g)$ for some $g \in T$. This result is proved for $\mathbb{B}$ in [161] and [490].

In the context of studying axiomatizations for $\mathbb{C}_{\exp}$ and related structures, Zilber was led to a stronger form of SC, called the Uniform Schanuel Conjecture, that we now describe. SC itself may be equivalently formulated as follows.

Let $V \subset \mathbb{C}^n \times (\mathbb{C}^\times)^n$ be a subvariety defined over $\overline{\mathbb{Q}}$ and of dimension $\dim V < n$. *Suppose $(z, e^z) = (z_1, \dots, z_n, e^{z_1}, \dots, e^{z_n}) \in V$. Then, $z_1, \dots, z_n$ are linearly dependent over $\mathbb{Q}$.*

Can something more be said about the set of linear dependence relations that arise in this way for a given V? Such a version might lend itself to first-order axiomatization, for example, if the set of $\mathbb{Q}$-linear dependencies is finite for any given V. This however fails. For example, consider

$$V \subset \mathbb{C}^3 \times (\mathbb{C}^\times)^3 : \quad z_1 z_2 = z_3^2, \quad w_1 = w_2 = w_3 = 1.$$

Thus, $\dim V = 2$. If $k_1, k_2, k_3 \in \mathbb{Z}$ with $k_1 k_2 = k_3^2$ and $z_i = 2\pi i k_i$, then $(z, e^z) \in V$, but these points do not lie in finitely many rational subspaces.

On the other hand, positing that, in the above situation, the e^z lie in finitely many proper subtori would give a statement weaker than SC. However, if e^z lies in a torsion coset of codimension 2 (or more), then this does imply a rational linear dependency on z (with two or more equations one can eliminate $2\pi i$). Therefore, the following statement implies SC.

18.13 Conjecture (Uniform Schanuel conjecture (USC); [550]) *Let $V \subset \mathbb{C}^n \times (\mathbb{C}^\times)^n$ be a subvariety defined over $\overline{\mathbb{Q}}$ and of dimension $\dim V < n$. Then there exists a finite set $\mathcal{L}$ of proper rational subspaces $L \subset \mathbb{C}^n$ with the following property. If $(z, e^z) \in V$, where $z = (z_1, \ldots, z_n)$, then there exists $L \in \mathcal{L}$ and $k = (k_1, \ldots, k_n) \in \mathbb{Z}^n$ such that $z + k \in L$. Moreover, if L has codimension 1 in $\mathbb{C}^n$, then $k = 0$.*

Note that USC, if true, is part of the first-order theory of $\mathbb{C}_{\exp}$. In the presence of the Zilber–Pink conjecture (which is a purely diophantine conjecture), SC implies USC ([550]). The key observation is the following.

Suppose V as above and $(z, e^z) \in V$. Let W be the image of V under projection to $(\mathbb{C}^\times)^n$. Let L be the smallest rational linear subspace containing z and $L = \exp(T)$ its image. Suppose that (z, e^z) lies in the Zariski open subset of V where the projection has generic dimension $\dim V - \dim W$. Let $A \subset_{\text{cpt}} W \cap T$ be the component containing e^z. By Schanuel's conjecture, tr. $\deg._{\mathbb{Q}}(z, e^z) \geq \dim T$, and (z, e^z) lies in a component of W over A, defined over $\overline{\mathbb{Q}}$, of dimension $\dim A + \dim V - \dim W$. So we have

$$\dim T \leq \dim A + (\dim V - \dim W) < \dim A + n - \dim W$$

(as $\dim V < n$), hence

$$\dim A > \dim W + \dim T - n$$

and thus A is an atypical intersection of W with the subtorus T.

If, as Zilber–Pink predicts, all atypical intersections of W with special subvarieties are accounted for by finitely many such intersections, then (assuming SC), and with a little extra work, one deduces USC.

18.14 Theorem ([550]) *If the Zilber–Pink conjecture holds in the multiplicative setting, then SC implies USC.*

Over the real numbers, SC implies USC; see [306]. On the model theory of $\mathbb{C}_{\exp}$, see further [307]. For some existential closedness results for abelian varieties, see [19]; for the j-function, see [18, 210], and for Shimura varieties, see [211]. Differential existential closedness for j is established in [17], enabling an axiomatization of j-reducts of differentially closed fields.

For a "two-sorted" categoricity result for exponentiation, see [552], and for analogous results for Shimura varieties and relations with Open Image theorems, see [164] (refined in [209] and [554]).

Lacunary Polynomials and Atypical Intersections

In the course of his investigations on reducibility over $\mathbb{Q}$ of lacunary polynomials of the form

$$F(x^{n_1}, x^{n_2}, \ldots, x^{n_k}),$$

where $F \in \mathbb{Z}[x_1, \ldots, x_k]$, and $n_i \in \mathbb{N}$, Schinzel [472] proved a theorem containing the following result, which concerns intersections of a curve $V \subset \mathbb{G}_m^3$ with the union of all subtori of dimension 1.

18.15 Theorem (Schinzel [472, Theorem 1]) *Let $P, Q \in \mathbb{C}[x_1, x_2, x_3]$ having* $\gcd(P, Q) = 1$. *Then there exists a constant $c(P, Q)$ with the following property. Suppose $n = (n_1, n_2, n_3) \in \mathbb{Z}^3, \xi \in \mathbb{C}^\times$ with*

$$P(\xi^{n_1}, \xi^{n_2}, \xi^{n_3}) = Q(\xi^{n_1}, \xi^{n_2}, \xi^{n_3}) = 0,$$

then either ξ is a root of unity or there is vector $\gamma \in \mathbb{Z}^3$ with $0 < h(\gamma) < c_1(P, Q)$ and $\gamma \cdot n = 0$.

The result was generalized to codimension-2 subvarieties $V \subset \mathbb{G}_m^n$ intersecting the union of one-dimensional subtori in [543]. (The result is a bit more precise. It shows, for $V \subset \mathbb{G}_m^n$ of codimension 2 or more, that the intersection V with the union of all one-dimensional subtori is contained in finitely many tori that intersect V in codimension 1; the extension to all one-dimensional torsion cosets is in [90]. See the discussion there on pp. 6–7.)

Bombieri–Masser–Zannier [87] formulated the generalization to curves $V \subset \mathbb{G}_m^n$, not contained in any proper special subvariety, proving it for $V/\overline{\mathbb{Q}}$ under the stronger hypothesis (removed later in [364]) that V is not contained in any proper *weakly* special subvariety. They also proved bounded height in that setting (bounded height for the relevant points is also obtained in [543]).

19

Formulations

Pink formulated his conjecture in terms of unlikely intersections, but here we follow Zilber and Bombieri–Masser–Zannier and consider atypical intersections (which, in the terminology of [90], include both unlikely and *anomalous* intersections: see Definition 20.5).

The Zilber–Pink Conjecture

The setting is a mixed Shimura variety X and its collection $S = S_X$ of special subvarieties. It is a basic fact that an irreducible component of the intersection of two special subvarieties is again a special subvariety.

Note also that a special subvariety Y of a mixed Shimura variety X is again a mixed Shimura variety. Its special subvarieties are precisely the intersection components of special subvarieties of X with Y.

19.1 Definition   Let X be a mixed Shimura variety with its collection S of special subvarieties, and let $V \subset X$ be a subvariety. A subvariety $A \subset V$ is called *atypical* (for V in X) if $A \subset_{\text{cpt}} V \cap S$ for some $S \in S$ with

$$\dim A > \dim V + \dim S - \dim X.$$

The collection S is countably infinite, so there are potentially countably infinitely many atypical subvarieties. In fact there can be infinitely many, but the conjecture is that when this happens they are subsumed in a larger atypical intersection, and that finitely many (maximal) atypical intersections account for all of them.

19.2 Conjecture (Zilber–Pink conjecture (ZP) [90, §5; 445, Conjecture 5.1; 550, Conjecture 2])   *Let X be a mixed Shimura variety with its collection S of special subvarieties, and $V \subset X$. Then V contains only finitely many maximal atypical subvarieties.*

Note that if V is contained in a proper special subvariety of X, then V itself is atypical, and the conjecture holds trivially. However, for any $V \subset X$, one can consider the intersection of all special subvarieties containing V. This is a special subvariety, and it is the smallest special subvariety containing V.

19.3 Definition For a mixed Shimura variety V and $V \subset X$ we denote by $\langle V \rangle_S$ (or often just $\langle V \rangle$) the smallest special subvariety containing V. We denote by $\mathcal{W}$ the collection of weakly special subvarieties of X and denote by $\langle V \rangle_{\mathcal{W}}$ the smallest weakly special subvariety containing V.

Thus ZP, considered for all mixed Shimura varieties, makes a non-trivial assertion in that case as well.

Since special subvarieties are part of the geometric structure of a mixed Shimura variety, this conjecture has the nice feature that the diophantine conjectures now have an (almost) purely geometric source. Its arithmetic content comes from the $\mathbb{Q}$-group and congruence subgroup underlying the construction of the mixed Shimura variety.

Optimal Subvarieties

Following [444] we introduce the *defect* of a subvariety, which measures how far it is from being special.

19.4 Definition Given X, S as above and $V \subset X$, the *defect* of V is defined by (and denoted)

$$\delta(V) = \delta_S(V) = \dim\langle V \rangle_S - \dim V.$$

19.5 Definition Given $V \subset X$, a subvariety $A \subset V$ is called *optimal* for V if one cannot increase A inside V without increasing the defect. Equivalently, $A \subset V$ is optimal if $A \subset B \subset V$ and $\delta(B) \leq \delta(A)$ imply $A = B$.

For example, V itself is always an optimal subvariety of V, but any proper subvariety $A \subset V$, if optimal, must be atypical for V (as a subvariety of $\langle V \rangle$). For A must then have a smaller defect, so that

$$\dim\langle A \rangle - \dim A = \delta(A) < \delta(V) = \dim\langle V \rangle - \dim V,$$

which re-arranges to the condition

$$\dim A < \dim V + \dim\langle A \rangle - \dim\langle V \rangle$$

for A to be atypical. For the same reason, such A must be a component of the intersection $V \cap \langle A \rangle$, otherwise one could decrease the defect by enlarging A to a component.

19.6 Definition   Let X be a mixed Shimura variety and $V \subset X$.

1. The *optimal set* of V is the set Opt(V) of its optimal subvarieties.
2. If Opt(V) is a finite set of subvarieties, then we call the formal sum of these the *optimal cycle* of V, and denote it also Opt(V).

19.7 Conjecture (ZP, Optimal form)   *Let X be a mixed Shimura variety and $V \subset X$. Then* Opt(V) *is a finite set.*

19.8 Proof of Equivalence of Conjectures 19.2 and 19.7

See [264]. In fact, the equivalence holds for each ambient X, provided it holds for every special subvariety of X considered with its own collection of special subvarieties.

Assume ZP as in Conjecture 19.2 holds for all $Y \in \mathcal{S}_X$, and that $V \subset X$. We may assume that $\langle V \rangle = X$ (or replace X by $\langle V \rangle$). We show that Opt(V) is finite by induction on dim V. We can assume that A is a proper subvariety and hence, as observed above, atypical. By Conjecture 19.2, A is contained in one of finitely many maximal atypical subvarieties B of V. Then $A \in$ Opt(B) and since dim $B < $ dim V, the set Opt(B) is finite by induction.

In the other direction, suppose that Opt(V) is finite. We may assume that $\langle V \rangle = X$, otherwise V is atypical and then is clearly the unique maximal atypical subvariety. Let $A \subset V$ be atypical. Let $B \supseteq A$ be an optimal subvariety containing A with $\delta(B) \leq \delta(A)$. Then B is atypical. So every atypical A is contained in one of finitely many $B \in$ Opt(V), and we take the finitely many maximal elements.   □

One immediately recovers MMM and AO, as it is clearly atypical for V to contain a special subvariety. More generally, ZP implies the corresponding special point result (i.e. Generalized André–Oort) for any mixed Shimura variety.

19.9 Proposition   *Assume ZP. Let X be a mixed Shimura variety and $V \subset X$. Then V contains only finitely many maximal special subvarieties.*

Proof   The maximal special subvarieties $A \subset V$ are optimal subvarieties of V.
                                                                 □

Further, ZP for hypersurfaces is equivalent to the special-point statement.

19.10 Proposition   *Let X be a mixed Shimura variety and $V \subset X$ a hypersurface. Then ZP for V is equivalent to the statement that V contains only finitely many maximal special subvarieties.*

Proof   As dim $V = $ dim $X - 1$, any component $A \subset V \cap S$ has dimension at least dim $S - 1$, and can only be atypical if $S \subset V$.   □

Weakly Optimal Subvarieties

One can also consider the optimality property with respect to the collection of weakly special subvarieties. We need this notion later in Chapters 21 and 22. Let X be a mixed Shimura variety with its collection $\mathcal{W} = \mathcal{W}(X)$ of weakly special subvarieties.

19.11 Definition   Given $X, S, \mathcal{W}$ as above and $V \subset X$, the *weakly special defect* of V (in X) is defined by (and denoted)

$$\delta_{\mathcal{W}}(V) = \dim \langle V \rangle_{\mathcal{W}} - \dim V.$$

(This was called *geodesic defect* in [264].) A subvariety $A \subset V$ is called *weakly optimal* (or *geodesic optimal*) if one cannot increase it among subvarieties of V without increasing the weakly special defect.

19.12 Remark   This notion appeared in earlier work [446] in model theory, in semi-abelian settings, where it was termed *cd-maximal* ("co-dimension maximal").

Weakly Special Subvarieties: Abelian Varieties

In general, for an abelian variety $A = A_x$, the subvariety $\{x\} \times A_x \subset \mathcal{X}_g$ is not a special subvariety, but it is always a weakly special subvariety.

19.13 Definition   Let X be a mixed Shimura variety, and $Y \subset X$ a weakly special subvariety. We define a *special subvariety* of Y to be an irreducible component of $Y \cap S$ for some $S \in \mathcal{S}_X$. We denote the collection of these $\mathcal{S}(Y)$. The *weakly special subvarieties* of Y are defined similarly and denoted as $\mathcal{W}(Y)$.

In case Y is special, this coincides with the special subvarieties of Y as a mixed Shimura variety.

If one defines the special subvarieties of A (identified with $\{x\} \times A_x$) in this way, the resulting collection is precisely the collection of torsion cosets of A (this is established in [445, 5.7]; for the weakly special statement, see [444, 4.6]).

Pink shows that ZP for $\mathcal{X}_g$ implies the corresponding ZP statement for a fibre A_x with its collection of special subvarieties (hence, ZP formally implies MM).

19.14 Definition   Let X be a mixed Shimura variety. A point $x \in X$ is called *Hodge generic* if it is not contained in any proper special subvariety of X.

Suppose $S \subset \mathcal{A}_g$ is special. Then the family of abelian varieties over S is a special subvariety $A \subset \mathcal{X}_g$, and it is a flat family of abelian varieties. Conversely, every special subvariety of $\mathcal{X}_g$ is essentially of this form: it is a flat family of torsion cosets over a special subvariety $S \subset \mathcal{A}_g$ (see [227, Proposition 1.1]).

19.15 Proposition ([445, 5.7]) *Let $A \to S$ be the family of abelian varieties over a special subvariety $S \subset \mathcal{A}_g$, and let $x \in S$ be a Hodge generic point. Then ZP for subvarieties of A_x with respect to the ambient A is equivalent to ZP for the ambient A_x with its collection of special subvarieties as in Definition 19.13.*

Proof We paraphrase the proof in [445, 5.7] in the language of optimal subvarieties. Let $V \subset A_x$. Any special subvariety B_x of A_x containing V is the fibre of a family of torsion cosets $B \to T$ over some special $T \subset \mathcal{A}_g$ with $S \subset T$ (as x is Hodge generic in S), so we can assume $T = S$. The family is flat, so that $\dim B = \dim B_x + \dim S$. If B_x is the smallest special subvariety of A_x containing V, then B is the smallest special subvariety of A containing V, so $\delta_A(V) = \delta_{A_x}(V) + \dim S$, and the optimal subvarieties of V for the two ambients coincide.   □

In any mixed Shimura variety, weakly special subvarieties likewise occur in flat families over a special subvariety, and the total space is also special (see, e.g. [40]).

19.16 Proposition *Suppose ZP holds for a mixed Shimura variety X. Then it holds for any weakly special subvariety $Y \subset X$ with special subvarieties defined as above.*

Proof The same as above.   □

Pink's Conjecture

Pink's formulation ([445, 1.3]) of his conjecture only concerns unlikely intersections (i.e. where dimension plus codimension fall short of the ambient dimension) rather than atypical ones. The following notation, a variant of that in [445], is convenient.

19.17 Definition Let X be a mixed Shimura variety with its collection $\mathcal{S}$ of special points. We denote by $X^{[k]}$ the union (countably infinite in general) of special subvarieties of X of *codimension* at least k.

19.18 Remark Beware that a number of variants of this notation appear in the literature, sometimes k denotes the *dimension* rather than the *codimension*.

19.19 Conjecture (Pink's conjecture)   *Let X be a mixed Shimura variety and V ⊂ X a subvariety that is not contained in any proper special subvariety of X. Then,*

$$V \cap X^{[\dim V + 1]}$$

is not Zariski-dense in V.

19.20 Proposition   *ZP implies Pink's conjecture.*

Proof   Suppose *V* is not contained in any proper special subvariety. Then, apart from *V* itself, every optimal subvariety is an atypical subvariety. And every unlikely intersection component *A* is contained in some atypical optimal subvariety.   □

Pink's conjecture also implies ZP in a formal way in many settings, such as the multiplicative setting (see [92]) and also (in a similar way) the modular setting. But they do not seem to be *formally* equivalent in general. For example, if *A* is a simple abelian variety of dimension 3 and $X = A \times A$, then special subvarieties of *X* are of dimension 0, 3, or 6. If $V \subset X$ with $\dim V = 3$ and not special (and so not contained in any proper special subvariety), then Pink's conjecture (applied repeatedly) implies AO: that *V* contains only finitely many special points. But ZP asserts that there are only finitely many atypical intersections with special subvarieties of dimension 3. Such intersections need not contain special points, and so ZP is, on its face, a stronger assertion. One can take *A* above to have CM, so that it is indeed a mixed Shimura variety, and one can also find examples among pure Shimura varieties (suitable Hilbert modular varieties).

Nevertheless, the two conjectures are equivalent, as shown by Barroero–Dill [39] (in view of [137, 233]). A different way of deducing the Zilber–Pink conjecture from Pink's is sketched in [432]; essentially, intersecting a given *V* with a suitably generic hyperplane turns atypical intersections into unlikely ones.

Curves

Consider in particular the case that *V* is a curve. Apart from *V* itself, the only possible optimal subvarieties are points, which are atypical if contained in a special subvariety of codimension at least 2. So ZP for a curve in a mixed Shimura variety *X* reduces to the following statement.

19.21 Proposition   *Let $V \subset X$ be a curve. Then ZP for V is equivalent to the statement that if V is not contained in a proper special subvariety, then $V \cap X^{[2]}$ is a finite set.*

Consider now the case of a curve $V \subset \mathbb{G}_m^2$, not contained in any proper coset. Suppose that $c_1, \ldots, c_k \in \mathbb{C}^\times$ are multiplicatively independent and let $\mathrm{div}\langle c_1, \ldots, c_k \rangle$ be the division group of the subgroup of $\mathbb{G}_m$ they generate. Consider the curve $V^* \subset \mathbb{G}_m^{n+k}$ given by setting

$$V^* = \{(x, y, c_1, \ldots, c_n) : (x, y) \in V\}.$$

A point (x, y, c) lies in $V^* \cap \left(\mathbb{G}_m^{2+k} \right)^{[2]}$ if and only if $x, y \in \mathrm{div}\langle c_1, \ldots, c_k \rangle$.

Hence, ZP implies Multiplicative Mordell–Lang for such curves (and similarly for any curve in $\mathbb{G}_m^n$).

On the other hand, taking an example in the style of [87], consider the curve $V \subset \mathbb{G}_m^{5+k}$ given parametrically by, for example,

$$V = \{(t, 1 - t, t + 3, 2t + 1, t^2 + 1, c_1, \ldots, c_k) \in \mathbb{C}^{5+k} : t \in \mathbb{C}\} \cap \mathbb{G}_m^{5+k},$$

with $c_1, \ldots, c_k$ multiplicatively independent as above. A point

$$(x, c) \in V \cap \left(\mathbb{G}_m^{5+k} \right)^{[2]}$$

could result from two coordinates x_i, x_j lying in $\mathrm{div}\langle c_1, \ldots, c_k \rangle$, and the finiteness of such points is governed by MML for the projection of V to those coordinates, but such a point could also result from two multiplicative dependences among the non-constant coordinates.

One sees that the assertion of ZP (which follows from the results of [87] in this case) is considerably stronger than ML.

Mordell–Lang

The above argument can be generalized to show that ZP for $\mathbb{G}_m^n$ (including for all its special subvarieties) implies MML. If Γ is a subgroup of $\mathbb{G}_m$, then a Γ-special point means a point with all coordinates in Γ, and a weakly special subvariety of $\mathbb{G}_m^n$ containing a Γ-special point is called a Γ-special subvariety.

19.22 Theorem *ZP implies MML.*

Proof Let $V \subset \mathbb{G}_m^n$ and let Γ be the division group of a finitely generated group, say the group generated by $c_1, \ldots, c_k \in \mathbb{C}^\times$, which we may assume to be multiplicatively independent. Put $c = (c_1, \ldots, c_k)$.

Suppose that $T = \langle V^* \rangle \subset \mathbb{G}_m^{n+k}$ is the smallest torsion coset containing V^*. Since $c_1, \ldots, c_k$ are multiplicatively independent, we have that T projects onto the k last coordinates with the fibre being of some dimension ℓ, $0 \leq \ell \leq n$. If V^* equals the fibre of T over $c \in \mathbb{G}_m^k$, then V is a Γ-special subvariety, and we have nothing further to prove.

Otherwise, it has dimension at most $\ell - 1$, while a Γ-point $x \in V$ gives rise to an intersection of V^* with a torsion coset of T of codimension ℓ. Such a point is then atypical. By ZP, V^* has finitely many optimal subvarieties that, apart from V^* itself, are atypical and contain all atypical points of V^*. So we have $(x, c) \in A^*$ for one of finitely many atypical $A^* = A \times \{c\} \in \mathrm{Opt}(V^*)$. Each such A^* is a component of $V^* \cap S$ for some torsion coset S, which must also be dominant to $\mathbb{G}_m^k$. If A^* is a fibre of $S \to \mathbb{G}_m^k$, then A is a Γ-special subvariety of V; otherwise it is a proper subvariety of the fibre and we can repeat the argument. □

More generally, Pink [445] and Zilber [550] show that their conjectures imply ML for (semi-)abelian varieties.

19.23 Theorem *Pink's conjecture (for the universal family of abelian varieties) implies ML for every abelian variety.*

Indeed, ZP implies the André–Pink–Zannier conjecture, which also implies semi-abelian ML.

20

Some Results

In this chapter, we state a selection of results giving special cases of ZP. The main objective is to illustrate the many different kinds of statements comprehended within the conjecture. There is a large literature, and while we have tried to give examples across the range of ZP, we have not attempted a complete catalogue. It would likewise take us too far afield to go into the methods of proof. Some results are classical, and a wide range of techniques have been employed; in addition to the specific references below (and earlier), see the surveys [128, 470]. The point-counting strategy has been employed in many recent results, especially (but not only) in those outside of the pure group settings. A second objective is to briefly describe the Poincaré bi-extension in order to see how the multiplicative group arises as a mixed Shimura variety.

Special Point Problems

ZP for any mixed Shimura variety implies the corresponding special point problem (Proposition 19.9). The Manin–Mumford and André–Oort conjectures have already been discussed in detail.

The Generalized André–Oort conjecture for the Legendre surface $\mathcal{L}$ was proved by André [7] (see also [425]). This was generalized to powers of $\mathcal{L}$ in [429].

20.1 Theorem ([429]) *Let $V \subset \mathcal{L}^n$. Then V contains only finitely many maximal special subvarieties.*

The special subvarieties of $\mathcal{L}^n$ are torsion sections over special subvarieties of $Y(2)^n$ and are described in detail in [429]. In fact, a more general result is proved there, combining powers of $\mathcal{L}$ with products of elliptic curves and copies of $\mathbb{G}_m$. The case in which V is a curve (in an elliptic surface) is proved effectively in [532]. The results of [429] have some overlap with those of [256], which characterizes subvarieties with a Zariski-dense set of special points in fibred powers of general elliptic surfaces.

Gao [225, 226] proves the special point result for a mixed Shimura variety assuming a Galois bound for special points in the underlying pure Shimura variety. In view of the subsequent results of [79, 433, 506], these hold in general.

20.2 Theorem ([225, 226, 433]) *The Generalized André–Oort conjecture holds in general.*

The Multiplicative Setting

Turning to results for atypical intersections, we consider first results in the basic setting of multiplicative groups. Bombieri–Masser–Zannier [87] prove that

$$V \cap \left(\mathbb{G}_m^n\right)^{[2]}$$

is finite for a curve $V \subset \mathbb{G}_m^n$ defined over $\overline{\mathbb{Q}}$ and not contained in any *proper weakly special subvariety* (an alternate proof is in [118] and a version for curves in $\mathbb{G}_m^n \times E^m$, where E is an elliptic curve, is in [38]), and suggest that the result should hold under the weaker hypothesis that V be not contained in any *proper special subvariety*. This strengthening was obtained by them ([89]) for curves in $\mathbb{G}_m^n$ up to $n = 5$, and then fully by Maurin [364].

20.3 Theorem (Maurin's Theorem; [364]) *Let $V \subset \mathbb{G}_m^n$ be a curve defined over $\overline{\mathbb{Q}}$, not contained in a proper special subvariety of $\mathbb{G}_m^n$. Then, $V \cap \left(\mathbb{G}_m^n\right)^{[2]}$ is a finite set.*

An alternative proof is given in [86]. At the other extreme, one has the result for codimension-2 subvarieties defined over $\overline{\mathbb{Q}}$ (finiteness of intersections with the union of one-dimensional special subvarieties).

20.4 Theorem ([90, Theorem 1.7]) *Let $V \subset \mathbb{G}_m^n$ defined over $\overline{\mathbb{Q}}$ with $\dim V = n - 2$. Then ZP holds for V.*

It is shown in [92] that results for subvarieties $V \subset \mathbb{G}_m^{r+s}$, $\dim V = r$ valid on a rectangle of dimension and codimension ($r \leq R, s \leq S$) extend from $\overline{\mathbb{Q}}$ to $\mathbb{C}$ on the same rectangle (for curves the reduction from $\mathbb{C}$ to $\overline{\mathbb{Q}}$ is in [88]). Hence, Theorems 20.3 and 20.4 hold for $V/\mathbb{C}$.

As ZP for hypersurfaces in $\mathbb{G}_m^n$ reduces to MMM, it follows from the above that ZP holds for all subvarieties (over $\mathbb{C}$) of $\mathbb{G}_m^n$ for $n \leq 4$. At present, the general case of surfaces in $\mathbb{G}_m^5$ is open, though there are some results [91] for planes (defined over $\overline{\mathbb{Q}}$).

20.5 Definition   Let $V \subset X$. A subvariety $A \subset V$ is called *anomalous* if $\dim A > 0$ and there is a weakly special subvariety U such that

$$\dim A > \dim V + \dim U - n.$$

The *open anomalous* set V^{oa} is the complement in V of the union of all its anomalous subvarieties, which we call the *anomalous locus* of V.

20.6 Theorem ([90])   *The open anomalous set V^{oa} is open in V.*

A key role in these ZP results is played by the Bounded Height conjecture formulated in [90, p. 5], with partial results, and proved in general by Habegger in [252] (see also an effective version in [257]).

If $V \subset \mathbb{G}_m^n$ is a curve, then the only possible anomalous subvariety is V itself, and so V^{oa} is either empty (if $V \subset U$ for a proper weakly special) or all of V. In the latter case, $V \cap S$ is a finite set for any proper special subvariety S, and it is shown in [87] that these points have bounded height. Such intersections are just likely intersections, and they are Zariski-dense in V.

If V is anomalous, then the heights of just likely intersections are in general not bounded. Situations where it must fail, and where it might not fail, are described in [90, §5].

20.7 Theorem ([252, 1.3]; formerly Bounded Height conjecture)   *Let $V \subset \mathbb{G}_m^n$ be defined over $\overline{\mathbb{Q}}$, and of dimension d. Then, $V^{\mathrm{oa}} \cap \left(\mathbb{G}_m^n \right)^{[d]}$ is a set of bounded height.*

In fact, bounded height is established for points of V near (in a height sense) subgroups of complementary dimension. (See [251] for ϵ-strengthenings of ZP in the multiplicative setting, and [365] for extension to include a finite rank subgroup for subgroups of codimension at least $\dim V + 1$; this is the analogue of a result [455] in an abelian variety.) Theorem 20.7 enables the following general result.

20.8 Theorem ([252, 1.4])   *Let $V \subset \mathbb{G}_m^n$ be a subvariety of dimension d defined over $\overline{\mathbb{Q}}$. Then,*

$$V^{\mathrm{oa}} \cap \left(\mathbb{G}_m^n \right)^{[d+1]}$$

is a finite set.

A subvariety $V \subset \mathbb{G}_m^n$ is called *degenerate* if $V^{\mathrm{oa}} = V$. The above methods are inapplicable to such subvarieties, but one has the following result of Bays–Habegger [44].

Suppose $V \subset \mathbb{G}_m^n$ is a curve defined over a number-field F. If $\sigma : F \to \mathbb{C}$ is a field embedding, then V^σ denotes the corresponding curve over $\mathbb{C}$. Let

$$\mathbf{T} = \{(x_1, \ldots, x_n) \in \mathbb{G}_m^n : |x_1| = \cdots = |x_n| = 1\}$$

be the maximal compact subgroup.

20.9 Theorem (Bays–Habegger [44]) *Let $n \geq 3$ and let $V \subset \mathbb{G}_m^n$ be a curve defined over a number-field F and not contained in any proper special subvariety. Suppose that $V^\sigma \cap \mathbf{T}$ is finite for some embedding $\sigma : F \to \mathbb{C}$. Then there are only finitely many points $x \in V$ with $x^n \in V$ for some $n \geq 2$.*

A point x, as in the theorem, gives rise to a point $(x, x^n) \in V \times V \subset \mathbb{G}_m^{2n}$ that is unlikely as such a point is in a special subvariety of codimension n, while $\dim(V \times V) = 2 < n$. The variety $V \times V$ is degenerate since, for any point $(x, y) \in V \times V$, we have $(x, y) \in \{x\} \times V \subset P \times \mathbb{G}_m^n$, a weakly special subvariety of codimension n intersecting $V \times V$ in a component of dimension 1. In addition to point-counting the proof uses linear forms in logarithms. There is also a variant for S-integral points using a bound for linear forms in p-adic logarithms.

Fewnomial theory is employed in [150] to obtain some finiteness results for the intersection of curves in $\mathbb{G}_m^n$ with real analytic subgroups.

Abelian Varieties

The Bounded Height conjecture in this setting is proved in [253]. Earlier cases under various hypotheses were in [523] and [453], the latter in particular settled the case of curves in an arbitrary abelian variety. The analogue of Theorem 20.6 is proved in [455].

The analogue of Maurin's Theorem for abelian varieties is established in [264], via point-counting. The main advantage of point-counting over previous approaches is that it does not require delicate height lower bounds in Lehmer-type problems. Earlier results had been restricted either to abelian varieties with CM ([119, 120, 450, 456, 523]) or products of elliptic curves ([224, 524]) or had other, stronger, hypotheses ([454, 455]). The result is extended to semi-abelian varieties by Barroero–Kühne–Schmidt in [41], relying on the semi-abelian Bounded Height conjecture established in [319].

20.10 Theorem (Barroero-Kühne-Schmidt [41]) *Let $V \subset X$ be a curve in a semi-abelian variety, both defined over $\overline{\mathbb{Q}}$. If V is not contained in a proper special subvariety, then $V \cap X^{[2]}$ is a finite set.*

The result extends to curves over $\mathbb{C}$ (in $G/\overline{\mathbb{Q}}$) by [40].

A variety of other results are established in [264] for subvarieties of abelian varieties defined over $\overline{\mathbb{Q}}$, including the analogue of Theorem 20.8 above, and a full conditional proof of ZP for an abelian variety under a suitable Galois-orbit conjecture. These (and also several previous results) consider the case of intersections with translates of abelian varieties by finite-rank subgroups.

For effective results under various additional hypotheses see, for example, [132], which gives an analogue of the result of [90] for codimension-2 subvarieties of $\mathbb{G}_m^n$ for products of CM elliptic curves. For products of non-CM elliptic curves, see [286]. These enable effective resolution of the Mordell conjecture in some circumstances; see also [133]. See further [134, 525].

It is shown by Barroero–Dill [39] that results for abelian varieties over $\overline{\mathbb{Q}}$ extend to abelian varieties defined over $\mathbb{C}$, so in particular Theorem 20.10 is known for a curve $V \subset A$, where A is an abelian variety, in full generality. Thus, ZP holds for an abelian variety $A/\mathbb{C}$ with $\dim A \leq 3$.

Relative Manin–Mumford

Pink showed that his general conjecture implies the following relative version of Manin–Mumford for abelian schemes.

20.11 Conjecture (Relative Manin–Mumford; RMM) *Let $A \to B$ be an algebraic family of abelian varieties over a base B, and $V \subset A$ a subvariety of dimension d that is not contained in any proper closed subgroup scheme of $A \to B$. Then, $V \cap A^{[>d]}$ is not Zariski-dense in V.*

Masser and Zannier have proved special cases of this in a series of papers starting with the following basic result on torsion anomalous points.

20.12 Theorem ([357]) *There are only finitely many complex numbers $\lambda \in Y(2)$ such that the points*

$$P_\lambda = (2, \sqrt{2(2 - \lambda)}), \quad and \quad Q_\lambda = (3, \sqrt{6(3 - \lambda)})$$

in the elliptic curve $E_\lambda : y^2 = x(x - 1)(x - \lambda)$ are both torsion.

20.13 Definition Let $\mathcal{L}$ be the Legendre family of elliptic curves. The *fibre powers* of $\mathcal{L}$ are defined by

$$\mathcal{L}^{(n)} = \{(\lambda, x_1, y_1, \ldots, x_n, y_n) : \lambda \in Y(2), (x_i, y_i) \in E_\lambda, i = 1, \ldots, n\}.$$

It is the family of nth powers E_λ^n over the modular curve $Y(2)$.

The locus of points

$$V = \{(\lambda, 2, \sqrt{2(2 - \lambda)}, 3, \sqrt{(6(3 - \lambda)}) : \lambda \in Y(2)\}$$

is a one-dimensional subvariety of $\mathcal{L}^{(2)}$ that has dim $\mathcal{L}^{(2)} = 3$. As the torsion sections

$$T_n = \{(\lambda, x_1, y_1, x_2, y_2) \in \mathcal{L}^{(2)} : \lambda \in Y(2), [n](x_1, y_1) = 0_{E_\lambda} = [n](x_2, y_2)\}$$

are one-dimensional special subvarieties, these torsion anomalous points in V are indeed atypical intersections. This is a relative Manin–Mumford statement as V is a subvariety of a moving family of abelian varieties. If V is contained in a single fibre, then the statement reduces to a case of the classical MM. For a variant for the Weierstrass family, see [255].

The basic result is generalized in [358] to a curve in an abelian surface scheme that is isogenous to the fibred product of two isogenous elliptic curve schemes. In particular, this handles the general case of a curve $V \subset \mathcal{L}^{(2)}$.

More general cases are dealt with in [360, 361], giving a general result for curves in abelian surface schemes over $\overline{\mathbb{Q}}$ (the earlier papers handle the case of non-simple abelian surface schemes over $\mathbb{C}$).

20.14 Theorem ([361]) *Let $\mathcal{A}$ be an abelian surface scheme over a variety defined over $\overline{\mathbb{Q}}$ and let V be an irreducible closed curve in $\mathcal{A}$. Then $V \cap \mathcal{A}^{[2]}$ is contained in a finite union of abelian subschemes of positive codimension.*

The full result confirming Conjecture 20.11 for a curve in an abelian scheme over $\mathbb{C}$ is achieved by Corvaja–Masser–Zannier [159], but now obtaining a result for a general subvariety. Let $\mathcal{A} \to S$ be an abelian scheme over a complex irreducible quasi-projective variety S, and $\sigma : S \to \mathcal{A}$ a section. An irreducible hypersurface $T \subset S$ is *torsion* for σ if the restriction of σ to T is torsion; that is, if there exists an integer n such that $n\sigma(s) = 0$ in the fibre $\mathcal{A}_s$ for all $s \in T$.

20.15 Theorem ([159]) *Let $\mathcal{A} \to S$ be an abelian surface scheme over a complex irreducible quasi-projective variety S. Let $\sigma : S \to \mathcal{A}$ be a section. Then there exists only finitely many torsion hypersurfaces in S, unless $\sigma(S)$ is contained in a proper subgroup scheme.*

See the above papers and also [363, 545] for applications to Pell's equation in polynomials, and problems concerning integration in elementary terms. See further [159] for density results for likely intersections in this context and connections with the *Betti map*, pursued further in [10, 230].

The generalization of Theorem 20.12 in which $(2, 3)$ is replaced by any pair $(\alpha, \beta) \in Y(2)(\mathbb{C})$ is considered in [359], where it is shown that the exceptional

set $T(\alpha,\beta)$ is finite provided $\alpha \neq \beta$, with precise effective results when
tr.deg.$_{\mathbb{Q}}(\alpha,\beta) = 1$. Another proof of this generalization of Theorem 20.12,
via dynamics is given by DeMarco–Wang–Ye [176], yielding strengthenings
related to the Bogomolov conjecture. Mavraki [366] (using dynamical methods) shows that $T(\alpha,\beta)$ is in fact empty in many algebraic cases, in particular
for the original case of $(\alpha,\beta) = (2,3)$. The results strengthen those of Stoll
[497], who also proves that there are only countably many pairs (α,β) with
$\#T(\alpha,\beta) \geq 2$ and conjectures that there are only finitely many pairs (α,β) for
which $\#T(\alpha,\beta) \geq 3$.

Note that Theorem 20.14 does not affirm ZP for curves in $\mathcal{L}^{(2)}$, as there are
other special subvarieties besides torsion sections. For a curve, one must also
consider atypical intersections where the parameter $x \in Y(2)$ is special and the
two points in E_λ are linearly dependent over the endomorphism ring of E_λ.
This is dealt with (in a more general situation) by Barroero in [34], proving the
following.

20.16 Theorem (Barroero [34])   *Let E_λ be the Legendre family of elliptic
curves, and $P_1,\ldots,P_n \in E_\lambda\big(\overline{\mathbb{Q}(\lambda)}\big)$ points that are linearly independent over
$\mathbb{Z}$. Then there are only finitely many complex $\lambda_0 \in Y(2)$ such that E_{λ_0} is CM
and $P_1(\lambda_0),\ldots,P_n(\lambda_0) \in E_{\lambda_0}$ are linearly dependent over $\mathrm{End}(E_{\lambda_0})$.*

This takes care of all possible atypical points on a curve $V \subset \mathcal{L}^{(2)}$, while
for two-dimensional subvarieties ZP reduces to the special-point problem and
follows from Theorem 20.1.

20.17 Theorem ([34]+ [429] + [358])   *ZP holds for the threefold $\mathcal{L}^{(2)}$.*

It was asserted in [445] that ZP implies the generalization of Conjecture 20.11
to *semi-abelian* families. However, Conjecture 20.11 for semi-abelian families
(and the implication) turn out to be *incorrect*, as observed by Bertrand [53].
The problem is that in the family of $\mathbb{G}_m$-extensions of a CM elliptic curve,
there are special subvarieties that contain infinitely many torsion points but
are not subgroup schemes. This family is known as the Poincaré bi-extension
(see Definition 20.21), and these special subvarieties are called *Ribet sections*.
Thus RMM, as stated in analogy with Conjecture 20.11, is false in general
for semi-abelian schemes, though ZP itself is affirmed in the counterexamples:
Bertrand–Masser–Pillay–Zannier [55] show that Ribet sections are the only
obstruction to the generalization of Conjecture 20.11 to semi-abelian surface
schemes over a curve.

20.18 Theorem ([55])   *Let E/S be an elliptic scheme over a curve $S/\overline{\mathbb{Q}}$, and
let G/S be an extension of E/S by $\mathbb{G}_m/S$. Furthermore, let $s: S \to G$ be a
section of G/S with image $W = s(S)$.*

Assume W meets infinitely many fibres in a torsion point of the fibre. Then either

(i) *s is a Ribet section or*
(ii) *s factors through a strict subgroup scheme of G/S.*

This shows that a full understanding of RMM for semi-abelian schemes requires the wider context of mixed Shimura varieties. Similarly, we see below that non-PEL special subvarieties are essential to understanding endomorphism rings of abelian varieties in a family. By contrast, RMM for abelian schemes involves only abelian subgroup schemes.

For a different type of result in a relative context, see [243].

Relative Unlikely Intersections

The above results represent relative forms of Manin–Mumford. Relative forms of unlikely intersections are also predicted by Conjecture 20.11, starting with statements for a curve in a family of abelian varieties; that is, relative forms of Theorem 20.10.

In the theorem below we consider a curve $V \subset \mathcal{L}^{(n)}$ that is dominant to the base $Y(2)$. Thus, λ is non-constant and may be used to uniformize V. A point on V may then be written as $(\lambda, P_1(\lambda), \ldots, P_n(\lambda))$, where $P_i(\lambda) \in E_\lambda$.

20.19 Theorem (Barroero–Capuano [35]) *Let $V \subset \mathcal{L}^{(n)}$ be an irreducible curve defined over $\overline{\mathbb{Q}}$, dominant to $Y(2)$. Suppose that the points $P_1(\lambda), \ldots, P_n(\lambda)$ are, generically, linearly independent over $\mathbb{Z}$. Then there exist only finitely many $\lambda_0 \in Y(2)$ for which $P_1(\lambda_0), \ldots, P_n(\lambda_0)$ satisfy two independent relations over $\mathbb{Z}$.*

The case of a CM value λ_0 and one linear relation over $\mathrm{End}\,(E_{\lambda_0})$ is again dealt with by Theorem 20.15, and this establishes ZP for curves defined over $\overline{\mathbb{Q}}$ in $\mathcal{L}^{(n)}$. The paper [37] of Barroero–Capuano considers curves in general abelian schemes.

20.20 Theorem (Barroero–Capuano [37]) *Let $\mathcal{A} \to S$ be an abelian scheme over a smooth irreducible curve S, and $V \subset \mathcal{A}$ an irreducible curve, all defined over $\overline{\mathbb{Q}}$. Suppose that V is not contained in a proper subgroup scheme of $\mathcal{A}$, even after a finite base change. Then the intersection of V with the union of flat subgroup schemes of $\mathcal{A}$ of codimension at least 2 is a finite set.*

This theorem does not yet affirm ZP for a curve in such $\mathcal{A}$ as the intersections with other codimension-2 special subvarieties, dealt with by Theorem 20.15 in fibred products of elliptic schemes, have apparently not been dealt with in this situation.

The Poincaré Bi-Extension

After possibly passing to a finite cover, every mixed Shimura variety is a torus bundle over a polarizable abelian scheme over a (pure) Shimura variety (see Milne [373, VI, Theorem 1.6, p. 399]). The *Poincaré bi-extension*, though somewhat exotic, is the simplest example illustrating this general form. In particular, it shows how $\mathbb{G}_m$ arises as a mixed Shimura variety.

Let A be an abelian variety. By an extension of A by $\mathbb{G}_m$ we mean a quasi-projective group variety G with

$$1 \to \mathbb{G}_m \to G \to A \to 1$$

exact. Extensions of A by $\mathbb{G}_m$ are parameterized by the dual abelian variety $A^\vee$ ([372, 11.2, 11.3; 485]). They are realized by removing the 0-sections from the *Poincaré bundle* on $A \times A^\vee$, which is the universal family of line bundles of degree 0 on A. We get the *Poincaré bi-extension* $\mathcal{P}_A$ of $A \times A^\vee$, whose fibres over $A^\vee$ give all extensions of A by $\mathbb{G}_m$ ([81]).

20.21 Definition   The family of Poincaré bi-extensions $\mathcal{P}_\lambda$ of $E_\lambda \times E_\lambda^\vee$ as λ varies over $Y(2)$ gives a mixed Shimura variety $\mathcal{P}$ of total dimension 4, arising as a quotient of $\mathbb{H} \times \mathbb{C}^3$. If $\lambda \in Y(2)$ is a special point, then the fibre $\mathcal{P}_\lambda$ is a three-dimensional special subvariety of $\mathcal{P}$.

For a description of the Poincaré bi-extension *qua* mixed Shimura variety, see [54].

Let $\mathcal{P}_\lambda$ be the Poincaré bi-extension of a CM elliptic curve E_λ. The fibre of $\mathcal{P}_\lambda$ over the origin in $E_\lambda \times E_\lambda^\vee$ is a copy of $\mathbb{G}_m$. This shows how $\mathbb{G}_m$ arises as a mixed Shimura variety. For a description of the Ribet sections and Ribet points, see [52, 53].

20.22 Theorem (Bertrand–Schmidt [57])   *ZP holds for* $\mathcal{P}_\lambda$.

Though ZP is a unifying conjecture, verifying it often requires its separation into cases. Thus, for a subvariety $V \subset \mathcal{P}_\lambda$ with $\dim V = 2$, ZP reduces to the special point problem in this mixed setting proved in [226]. For curves $V \subset \mathcal{P}_\lambda$, torsion points on a section are considered in [55] (as in Theorem 20.18), whereas, in general, intersections with the Ribet section may be non-torsion, though in a rank-1 subgroup. If the curve lies in a fixed semi-abelian variety, then finiteness follows from semi-abelian Mordell–Lang. The "general" case is treated in [57].

The André–Pink–Zannier Conjecture

In a relative setting, this conjecture (18.11 above) encompasses certain relative Mordell–Lang problems. For a discussion (and results), see [179, 180, 181, 227, 335, 397]. We state the main result by Dill in [180].

Let A be an abelian scheme of relative dimension g over a curve S, and write $\pi : A \to S$ for the projection, and $A_s \subset A$ for the fibre over $s \in S$. Fix a point $0 \in S$ and a subgroup $\Gamma \subset A_0$. A point $p \in A$ lies in the *isogeny orbit* of Γ (in A) if there exists an isogeny $\psi : A_{\pi(p)} \to A_0$ with $\psi(p) \in \Gamma$. Let A_Γ denote the isogeny orbit of Γ.

If Γ is finite rank, then Conjecture 18.11 characterizes when a subvariety $V \subset A$ can have a dense set of points in the isogeny orbit of Γ. If $S \subset \mathcal{A}_g$ and A is the restriction of $\mathcal{X}_g \to \mathcal{A}_g$, then the conclusion would be that V is, in particular, weakly special.

20.23 Theorem (Dill [180]) *Let $A \to S$ be an abelian scheme of relative dimension g over a curve S, all defined over a number-field K. Let ξ be the generic point of S. Suppose that $A \to S$ is not isotrivial, and that, over $\overline{K(S)}$, A_ξ is isogenous to a power of an elliptic curve. Suppose further that A_0 is isogenous to E_0^g where $\mathrm{End}(E_0) = \mathbb{Z}$. Let $V \subset A$ be irreducible. If A_Γ is Zariski-dense in V, then one of the following two conditions holds:*

(i) *For some $s \in S$, V is a translate of an abelian subvariety of A_s by a point of $A_\Gamma \cap A_s$.*

(ii) *Over $\overline{K(S)}$, V_ξ is a union of translates of abelian subvarieties of A_ξ by torsion points of A_ξ.*

See also [181] for an approach using the existence of Galois automorphisms acting as homotheties.

Turning to pure settings, the modular case is a special case of a theorem established in [263] (when $U \subset \overline{\mathbb{Q}}$; the general case is in [431]) dubbed Modular Mordell–Lang.

20.24 Definition Let $U \subset \mathbb{C}$. A *U-special point* of $Y(1)^n$ is a point $(x_1, \ldots, x_n)$ such that each x_i is either special or in the Hecke-orbit of some $u \in U$. A *U-special subvariety* of $Y(1)^n$ is a weakly special subvariety containing a U-special point.

20.25 Theorem (Modular Mordell–Lang [263, 431]) *Let $V \subset Y(1)^n$. Then V contains only finitely many maximal U-special subvarieties.*

In $\mathcal{A}_g$, one has the following result of Orr [397].

20.26 Theorem (Orr [397]) *Let $V \subset \mathcal{A}_g$ be an irreducible curve and Λ an isogeny class in $\mathcal{A}_g$ such that $V \cap \Lambda$ is Zariski-dense in V. Then V is weakly special.*

For a general conditional result, see [459]; the result is made unconditional for arbitrary Shimura varieties of abelian type in [555]. Notwithstanding the ineffectiveness of these results, Freitag–Scanlon [221] show (answering a question in [368]) that if one considers the Hecke-orbit of a transcendental element, then the theory of differential fields can be used to derive effective and explicit bounds from finiteness.

20.27 Theorem ([221, §5.1]) *If $\psi : \mathbb{P}^1 \to \mathbb{P}^1$ is any non-identity automorphism of the projective line with graph V and $t \in \mathbb{A}^1(\mathbb{C}) \backslash \mathbb{A}^1(\overline{\mathbb{Q}})$, then the set of $s \in \mathbb{A}^1(\mathbb{C})$ for which E_s is isogenous to E_t and $E_{\psi(s)}$ is isogenous to $E_{\psi(t)}$ has cardinality at most $2^{38} \cdot 3^{14}$.*

More generally, consider a subvariety $V \subset Y(1)^n$ and an n-tuple $u = (u_1, \dots, u_n)$ of non-algebraic points. Let $\Theta \subset V$ be the set of $x = (x_1, \dots, x_n) \in V$ such that E_{x_i} is isogenous to E_{u_i} for $i = 1, \dots, n$. By Theorem 20.25 the Zariski closure of Θ is a finite union of weakly special subvarieties, and [221, 5.9] gives an explicit bound on its degree. More generally, they give bounds for any Kolchin-closed subvariety.

Returning to the mixed setting, results of [397] are extended by Gao in [227] to $\mathcal{X}_g$, in particular affirming the case in which the subvariety is a curve.

20.28 Theorem (Gao [227]) *Conjecture* 18.11 *holds for a curve $V \subset \mathcal{X}_g$.*

The André–Pink–Zannier for Galois generic points is established in [115]. In a different direction, topological density and equidistribution is established for more restrictive S-Hecke orbits, using methods of homogeneous dynamics, in [458].

Atypical Intersections in Pure Shimura Varieties

Various partial and conditional results in powers of $Y(1)$ are given in [263, 264]. These are discussed in detail in Chapters 21 and 22. In particular we describe a full proof of ZP for $Y(1)^n$ conditional on suitable Galois orbit bounds for unlikely intersections. For some different partial results, see [15].

This has been generalized by Daw–Ren [169], who show that ZP for a general Shimura variety follows from suitable arithmetic hypotheses, including Galois orbit bounds and others. These other hypotheses are affirmed for $\mathcal{A}_2$ in

[167, 168], which concern reduction theory. This gives a proof of ZP for $\mathcal{A}_2$ conditional only on suitable Galois orbit bounds.

Here we state an unconditional result similar in spirit to the result of [263] establishing ZP for *asymmetric* curves over $\overline{\mathbb{Q}}$ in $Y(1)^n$, described in Chapter 21. A curve $V \subset \mathcal{A}_g \times \mathcal{A}_g$ is called *asymmetric* if the degrees of the two projections to $\mathcal{A}_g$ are different (see [400] for the precise condition). Let Σ be the set of points $(s_1, s_2) \in \mathcal{A}_g \times \mathcal{A}_g$ such that the abelian variety A_{s_1} is isogenous to A_{s_2} and $\text{End}(A_{s_1}) = \mathbb{Z}$.

20.29 Theorem (Orr [400]) *Let $g \geq 2$. Let $V \subset \mathcal{A}_g \times \mathcal{A}_g$ be an asymmetric curve defined over $\overline{\mathbb{Q}}$. Suppose that V is not contained in a proper special subvariety of $\mathcal{A}_g \times \mathcal{A}_g$. Then $V \cap \Sigma$ is finite.*

Non-simple Abelian Varieties in a Family

Let $A \to S$ be a family of abelian varieties defined over a number-field k (or finite type field) whose generic fibre is simple. The set of points $x \in S(k)$ such that the fibre A_x fails to be simple is studied in [205]. Various results are proved to the effect that the exceptional set is sparse, using approaches from both arithmetic geometry and analytic number theory.

In several of the applications one even knows that generically $\text{End}(A_x) = \mathbb{Z}$, and then if the fibre fails to be simple, then it also acquires additional endomorphisms. This connects the problem with results in [349], where sparseness results are proved for the set of fibres in a family of abelian varieties for which the specialization acquires extra endomorphisms. Related results are in [4]. It would take us too far afield to discuss the methods, results, and questions in these papers; here we want to connect these problems with ZP and observe an interesting feature.

A family $A \to S$ of (principally polarized) abelian varieties of dimension g factors through a map to $\mathcal{A}_g$, and for our purposes we assume that we are considering the family over a subvariety $V \subset \mathcal{A}_g$, and for concreteness we further suppose that V is a curve, defined over $\mathbb{C}$.

Now being non-simple, or (more generally) having $\text{End}(A_x) \neq \mathbb{Z}$, are special conditions, and place x on a proper special subvariety. Moreover, these special subvarieties have quite high codimension. If A_x is not simple, then x lies on a special subvariety of dimension at most $\dim \mathcal{A}_{g-1} + 1$, hence codimension at least $g - 1$. If A_x is simple but $\text{End}(A) \neq \mathbb{Z}$, then (see [139], pp. 552–553) $\text{End}(A_x)$ contains either an order in a totally real field $K \neq \mathbb{Q}$, or an order in an imaginary quadratic field L. Shimura [488] computes the dimensions of the corresponding moduli spaces. The first case corresponds to Hilbert modular varieties, and the codimension is at least $g^2/4$ (achieved if g is even), the second case has codimension $(g^2 + 2g)/4$ or $(g^2 + 2g + 1)/4$ according to whether g is even or odd.

Conclusion: If $g \geq 2$, then being non-simple or having a non-trivial endomorphism entails lying on a special subvariety of codimension at least $g - 1$. We can then observe the following consequence.

20.30 Proposition *Assume $g \geq 3$ and that ZP holds for $\mathcal{A}_g$. Let $V \subset \mathcal{A}_g$ be a curve that is not contained in any proper special subvariety of $\mathcal{A}_g$. Then there are only finitely many points $x \in V$ (over $\mathbb{C}$) for which A_x has a non-trivial endomorphism.*

We can ask: if the conclusion fails, can we say something more about the proper weakly special subvariety that V is contained in? Must it be one corresponding to having an extra endomorphism, or even of failing to be simple?

In dimension $g \leq 3$, all special families of abelian varieties are determined by a Hodge class that defines a special endomorphism (see [185, 17.1]). In dimension 4, one finds the first examples of non-PEL special subvarieties, constructed by Mumford [382], and with it a negative answer to the above question. The special points on the Mumford–Shimura curves in $\mathcal{A}_4$ are analyzed in [390]. The generic abelian variety B in the family is simple with $\mathrm{End}(B) = \mathbb{Z}$, but there are infinitely many special points for which the corresponding abelian variety factors as a product of an elliptic curve and an abelian threefold.

We saw above a similar phenomenon with torsion anomalous points in families of semi-abelian varieties (and the absence of this phenomenon for families of abelian varieties). It seems interesting to ask in which cases one can refine ZP to say that if V has too many intersections with special subvarieties "of a given type", then V is contained in a proper special subvariety "of that type".

Independence of CM Points in Elliptic Curves

Suppose that X is a Shimura variety, and A is an abelian variety, both of dimension n, and suppose $V \subset X \times A$, with $\dim V = n$. Then $X \times A$ is (isomorphic to) a weakly special subvariety of $X \times \mathcal{A}_n$, and ZP predicts that it too satisfies ZP with respect to its special subvarieties, which are products $Y \times B$ of special $Y \subset X$ and torsion cosets $B \subset A$ (see [438]). Consider, for example, a point $(x, y) \in V$ such that x is special and y lies in a proper torsion coset B. Then (x, y) lies in the special subvariety $\{x\} \times B$ of codimension $n + n - \dim B > n$, and so such a point is atypical.

This may seem a rather esoteric case of ZP; however, it comes up in connection with some classical problems. Let E be an elliptic curve defined over $\mathbb{Q}$. By a famous theorem of Wiles et al., E has a parameterization

$$\phi: X_0(N) \to E$$

by modular functions, where N is the *conductor* of E and $X_0(N)$ is the compactification of $Y = Y_0(N)$. The images of special points are then referred to as *CM points* and (usually under some additional conditions) *Heegner points*. The trace of a Heegner point is then in $E(\mathbb{Q})$, and proving it is non-trivial under appropriate hypotheses is connected with the Birch–Swinnerton-Dyer conjecture and the deep work of Gross–Zagier and Kolyvagin. For this it is necessary (though not sufficient) that the CM point itself be non-torsion; see [541, pp. 76 and 79–81].

Let $V \subset Y \times E$ be the graph of such a parameterization. Special points on V correspond to special points $s \in Y$ whose image $\phi(s)$ is torsion in E. As V itself can never be special, ZP predicts that only finitely many CM points are torsion. This is affirmed for Heegner points under modular parameterizations in [387].

Going further, the linear independence of Heegner points corresponding to orders in distinct imaginary quadratic fields whose (odd part of the) class numbers are sufficiently large are is established in [463]. The results are strengthened (in some respects) in [320]. Both results use class field theory. More generally, let $V \subset Y \times E$ be a *correspondence* between an elliptic curve and a modular (or Shimura) curve. (That is, V is a curve dominant to both factors.) We call $x \in E$ a V-*image* of $s \in Y(1)$ if $(s,x) \in V$. Finiteness for the number of V-images of special points lying in a finite rank subgroup $\Gamma \subset E$ is proved in [110].

As we have seen, any linear relation among CM points is atypical. The question is studied in [438] within the ZP framework, obtaining further improvements in some respects. We do not compare the results in detail with those in [110, 320, 463] (see also [32]). The hypotheses vary (concerning whether $V \subset Y \times E$ is the graph of a modular parameterization or just an arbitrary correspondence, whether one restricts to Heegner points, whether linear independence is over $\mathbb{Q}$ or with respect to the full endomorphism ring of E, etc.), and also the conclusions are not all strictly comparable (e.g. [110] also get finiteness for points lying near Γ in a height sense).

Essentially, [438] affirms what ZP implies about linear independence of CM points. The paper considers a correspondence $V \subset Y \times E$, where E is an elliptic curve defined over $\mathbb{C}$, and Y may be a modular curve or a Shimura curve (as also in [110]). Here we restrict to $Y = Y(1)$. We allow E to have CM, in which case *linear independence* in E means with respect to $\mathrm{End}\,(E)$.

20.31 Definition   Let D be a positive integer. A set of special points $s_1, \ldots, s_n \in$ $Y(1)$ is called *D-independent* if, for each i, $\Delta(s_i) \geq D$ and, for $i \neq j$, there is no relation $\Phi_N(s_i, s_j) = 0$ with $N \leq D$.

Note that special points corresponding to different quadratic fields can never be connected by a modular relation.

20.32 Theorem (Pila–Tsimerman [438, Corollary 1.2])   *Let E be an elliptic curve defined over $\mathbb{C}$, $V \subset Y(1) \times E$ a correspondence, and $n \geq 1$. There exists $D = D(E, V, n)$ such that if $\{s_1, \ldots, s_n\}$ is D-independent, then any V-images $x_1, \ldots, x_n$ in E of $s_1, \ldots, s_n$ are linearly independent in E.*

20.33 Corollary ([438, Corollary 1.4])   *With the hyptheses as above, let Σ denote the set of V-images in E of special points in $Y(1)$. Let r be a non-negative integer. There exists $N = N(E, V, r)$ such that if $\Gamma \subset E$ is a subgroup of rank r, then $\#\Sigma \cap \Gamma \leq N$.*

Theorem 20.32 is proved by considering the corresponding atypical inter-section for

$$V^n \subset Y(1)^n \times E^n.$$

However, the methods are not able to confirm ZP for V^n once $n \geq 3$. The available arithmetic enables the control of linearly dependent CM points, but not other kinds of atypical intersections.

Multiplicative relations among singular moduli are studied in [437] (see also [217]). There also the methods can deal with some atypical intersections but not others. Results in a different direction concerning multiplicative relations with torsion points on a curve in $E \times \mathbb{G}_m^n$, where E is an elliptic curve, are obtained in [42].

Generalization of ZP to Variations of Hodge Structure

A generalization of ZP to the setting of variations of mixed Hodge structure over a smooth quasi-projective variety is set out in Klingler [308]. The formulations include Ax–Schanuel in this setting, which has been proved in [30, 137, 233]. See also [553].

Some Effective Results on Unlikely Intersections

Six problems of unlikely intersections in the setting of the Legendre family are considered by Habegger–Jones–Masser [262]. Four properties are considered for a fibre E_λ: E_λ is CM; $E_{-\lambda}$ is CM; $(2, \sqrt{4 - 2\lambda}) \in E_\lambda$ is torsion; and λ is a root of unity. For each pair of properties, there are only finitely many λ such that both properties hold, being examples of AO, Generalized AO, or Relative Manin–Mumford. But the general results are ineffective. In this paper, these cases are made effective, and in some cases explicit.

The special point problem in $Y(1) \times \mathbb{G}_m$ is dealt with effectively, via class field theory in [405], and via logarithmic forms in [404].

Additive Extensions

Additive extensions of semi-abelian varieties are not part of the mixed Shimura picture. However, it is natural to consider the corresponding problems, which also arise in the context of Pell's equation in polynomials.

20.34 Theorem (Schmidt [474]) *Let G be an extension by $\mathbb{G}_a$ of an elliptic scheme over a variety defined over $\mathbb{C}$ and denote by $G^{[c]}$ the union of its flat subgroup schemes of codimension at least c. Let $V \subset G$ be a curve. Then $V \cap G^{[2]}$ is contained in a finite union of subgroup schemes of G of positive codimension.*

A sharpening of the original Manin–Mumford conjecture for an extension of an elliptic curve by $\mathbb{G}_a$ is obtained in [158]. A Manin–Mumford Theorem for the additive extension of a product of elliptic curves is in [296]. A variant of Mordell–Lang in an additive extension of an abelian variety is studied in [238]. See also [228].

Analogues of ZP

ZP addresses atypical interaction between two collections of subvarieties of X: the collection of algebraic subvarieties defined over $\mathbb{C}$, and the collection of special subvarieties of X. The analogous problem when the special structure is enlarged to be the collection of algebraic varieties defined over $\overline{\mathbb{Q}}$ is studied in [131], obtaining the full analogue of ZP for atypical intersections of a variety $V \subset \mathbb{P}^n(\mathbb{C})$, defined over $\mathbb{C}$, with varieties defined over $\overline{\mathbb{Q}}$. The bounded height conjecture (of [131]) in this setting is established in [240].

Some partial results on a modular ZP with derivatives (interpolating between SC and USC in that setting) are obtained in [14, 494]. Different results related to ZP in a differential field are obtained in [432].

Unlikely intersections for curves in additive extensions in positive characteristic are studied in [106]. A dynamical analogue is studied in [241].

21

Curves in a Power of the Modular Curve

Statements and Results

In this chapter, we describe the application of the point-counting strategy to curves in $Y(1)^n$, leading to some partial results, and some conditional results. The main results (Theorems 21.6, 21.7, and implicitly also Theorem 21.3) are from Habegger–Pila [263]. We adapt the exposition there.

Let $V \subset Y(1)^n$ be a curve. The only subvarieties of V are points and V itself, and thus (as observed earlier) ZP reduces to the following statement in this case.

21.1 Conjecture *Suppose that $V \subset Y(1)^n$ is a curve defined over $\mathbb{C}$. Then $V \cap (Y(1)^n)^{[2]}$ is a finite set unless V is contained in a proper special subvariety of $Y(1)^n$.*

The point-counting strategy succeeds in proving this provided certain Galois orbits are large. Before framing various results, let us consider the kinds of points arising in $V \cap (Y(1)^n)^{[2]}$. A special subvariety of codimension 2 is defined by two special relations, each of one of two types:

(i) $x_i = \sigma_i$ for some singular modulus σ_i.

(ii) $\Phi_N(x_i, x_j) = 0$ for some $N \geq 1$ and $i \neq j$.

Suppose that $V \subset Y(1)^n$ is a curve that is not contained in any proper special subvariety. If both conditions defining the special subvariety S are of the form (i), say $x_i = \sigma_i, x_j = \sigma_j, i \neq j$, then a point in the intersection $V \cap S$ arises when the image V_{ij} of V under projection to the ij-plane contains the special point (σ_i, σ_j). Since V is not contained in any proper special subvariety, V_{ij} is not a special curve (or a special point). Therefore, by the André–Oort conjecture for curves in $Y(1)^2$ (theorem of André), for each pair ij there are only finitely many such points.

180

At the other extreme, a special subvariety S defined by two conditions of type (ii) is strongly special. Here we have modular relations between two pairs of coordinates

$$\Phi_N(x_i, x_j) = 0, \quad \Phi_M(x_k, x_\ell) = 0, \quad N, M \geq 1,$$

where $\{i, j\} \neq \{k, \ell\}$ (though the sets are permitted to intersect).

In between, a special subvariety may be defined by one equation of each type (i) and (ii). Such a special subvariety is not strongly special, and it is convenient to denote by

$$\left(Y(1)^n \right)^{[2], \text{nss}}$$

the union of all not strongly special subvarieties of $Y(1)^n$ of codimension at least 2.

The required Galois orbit conjecture is the following. A general form of this conjecture (formulated in the next chapter) enables a full proof of ZP for $Y(1)^n$.

21.2 Conjecture *Let V be as above defined over a field L of finite type. There are positive constants $c(V, L), \delta(V, L)$ such that for any intersection point $x = (x_1, \ldots, x_n)$ of V with a special subvariety T of codimension 2 we have*

$$[L(x) : L] \geq c\Delta(T)^\delta.$$

(Note that such x is always algebraic over L; also, if the conjecture holds for one field of definition of finite type for V, then it holds for any such field.)

Under this conjecture for a curve V one can verify Conjecture 21.1 for V. We state this but then describe various situations in which one can verify the conjecture and obtain some unconditional results.

21.3 Theorem (Essentially [263]) *Let $V \subset Y(1)^n$ be a curve. If Conjecture 21.2 holds for V, then ZP (i.e. Conjecture 21.1) holds for V.*

It turns out that Conjecture 21.2 always holds if $V \subset Y(1)^3$ is not defined over $\overline{\mathbb{Q}}$. This is because the modular curve defined by $\Phi_N(x, y) = 0$ has large *gonality* (minimal degree of a map from the curve to $\mathbb{P}^1$; it grows as a positive power of N). For the details of this, see [431].

21.4 Theorem ([431, Theorem 1.4]) *Let $V \subset Y(1)^3$ be a curve not defined over $\overline{\mathbb{Q}}$. Then ZP holds for V.*

For curves defined over $\overline{\mathbb{Q}}$, the Galois bounds are established in [263] for curves satisfying a certain (non-generic) condition.

21.5 Definition   A curve $V \subset Y(1)^n$ is called *asymmetric* if, in the sequence of degrees $\deg X_i|_V$ of the coordinate functions restricted to V, each positive integer appears at most once, up to one exception, which may appear at most twice.

21.6 Theorem ([263, Theorem 1])   *Suppose $V \subset Y(1)^n$ is an asymmetric curve defined over $\overline{\mathbb{Q}}$. Then ZP holds for V.*

Thus, in addition to a conditional result, and a result for curves not defined over $\overline{\mathbb{Q}}$ (which may be considered a generic result), we have a result for a non-generic class of curves over $\overline{\mathbb{Q}}$. Finally, the required Galois bounds can always be established for non-strongly special intersections, so we get an unconditional result when the atypical intersections are of a non-generic type.

21.7 Theorem ([263, Theorem 2])   *Suppose $V \subset Y(1)^n$ is a curve defined over $\overline{\mathbb{Q}}$, not contained in any proper special subvariety. Then $V \cap \left(Y(1)^n\right)^{[2],\text{nss}}$ is a finite set.*

Proof Strategy

According to the above discussion, we need to consider two kinds of special subvarieties and, for each, a finite number of choices of coordinates:

1. Two pairs of coordinates with each pair linked by a modular relation.
2. One coordinate takes a special value, two other coordinates are linked by a modular relation.

For each of these, there is a finite number of choices for the coordinates: let us consider an intersection point $x = (x_1, \ldots, x_n) \in Y(1)^n$ of V with a special subvariety of the first type defined by

$$\Phi_N(x_i, x_j) = 0, \quad \Phi_M(x_k, x_\ell) = 0, \quad N, M \geq 1,$$

where $\{i, j\} \neq \{k, \ell\}$ (though the sets are permitted to intersect).
For $\alpha, \beta \in \mathrm{GL}_2^+(\mathbb{R})$, we let

$$Y_{\alpha,\beta} = \{(z_1, \ldots, z_n) \in \mathbb{H}^n : z_j = \alpha z_i, z_\ell = \beta z_k\}.$$

This a definable (indeed semi-algebraic in real coordinates) set in $\mathbb{H}^n$, and a complex algebraic subvariety of $\mathbb{H}^n$ of dimension $n - 2$.

Let $Z = j^{-1}(V) \cap F^n$, where F is the standard fundamental domain for the j-function, and let $z_i, z_j, z_k, z_\ell \in Z$ be pre-images of x_i, x_j, x_k, x_ℓ. Then there exists $\alpha, \beta \in \mathrm{GL}_2^+(\mathbb{Q})$ such that

$$z_j = \alpha z_i, \quad z_\ell = \beta z_k,$$

and thus we get a rational point on the definable set

$$X = \{(\alpha, \beta) \in \mathrm{GL}_2^+(\mathbb{R}) \times \mathrm{GL}_2^+(\mathbb{R}) : Y_{\alpha,\beta} \cap Z \neq \emptyset\}.$$

We apply the Counting Theorem to this set. We must then estimate the height of the rational point coming from our point x above, and the next chapter proves the requisite (rather easy in this case) lemmas. We must also show that we get many points, and this is accomplished in Theorem 21.3 using the assumption of Conjecture 21.2. For the subsequent theorems it is achieved using (less easy) arithmetic estimates which we take up after the proof of Theorem 21.3 or, for Theorem 21.4, gonality results, as mentioned.

These ingredients enable us to prove that X contains positive-dimensional semi-algebraic sets accounting for the many rational points. There are now two issues to be dealt with which do not arise in the application of counting to special-point problems.

1. The set X is fibred by positive-dimensional semi-algebraic sets due to stabilizers: if $z \in Y_{\alpha,\beta} \cap Z$, then $z \in Y_{g_j \alpha g_i, h_\ell \beta h_k} \cap Z$ whenever $g_i z_i = z_i$ and $h_k z_k = z_k$. Thus, $X = X^{\mathrm{alg}}$ and to get a non-trivial conclusion from counting we must apply the more refined version in Theorem 9.14 (involving underalgebraic cells, or blocks).

2. Having done this, we must show that the semi-algebraic sets in X cannot stay inside the stabilizers, but that the intersection points must move along some semi-algebraic set obtained. This then leads via functional transcendence (here Modular Ax–Logarithms) to a contradiction of the hypothesis that V was not contained in a proper special subvariety. We do this with further arguments hinging on o-minimality properties. (Or one can count semi-rational points using Theorem 9.16, in which these arguments are incorporated.)

Height Estimates

The first lemma is a simple quantification of the fact that the standard fundamental domain is indeed a fundamental domain for the $\mathrm{SL}_2(\mathbb{Z})$ action on $\mathbb{H}$, and is easily proved by an elaboration of the proof of this as given, for example, in [484]. Set for $z \in \mathbb{H}$,

$$D(z) = \max\{1, \mathrm{Re}\,(z), \mathrm{Im}\,(z)^{-1}\}.$$

We let $\mathrm{Mat}_2(\mathbb{R})$ denote the set of 2×2 real matrices, which can be identified with $\mathbb{R}^4$ in the obvious way. The height $H(\alpha)$ of some $\alpha \in \mathrm{Mat}_2(\mathbb{Q})$ is defined accordingly.

The following is a variant of [427, Lemma 5.2] and [263, Lemma 5.1].

21.8 Lemma   *There exist absolute positive constants c, δ with the following property. Let $z \in \mathbb{H}$. Then there exists $\gamma \in \mathrm{SL}_2(\mathbb{Z})$ with $\gamma z \in F$ satisfying*

$$H(\gamma) \leq cD(z)^{\delta}.$$

Proof   If $\mathrm{Im}(z) \geq 1$, then we can take g to be of form $g = \begin{pmatrix} 1 & n \\ 0 & 1 \end{pmatrix}$ with $|n| \leq |\mathrm{Re}(z)| + 1$. So we may assume $\mathrm{Im}(z) < 1$. For $g = \begin{pmatrix} a & b \\ c & d \end{pmatrix} \in \mathrm{SL}_2(\mathbb{Z})$, we have

$$\mathrm{Im}(gz) = \frac{\mathrm{Im}(z)}{|cz + d|^2}.$$

Therefore, $\mathrm{Im}(gz)$ has a maximum as g varies over $\mathrm{SL}_2(\mathbb{Z})$, and it is attained for some g with $|c| \leq \mathrm{Im}(z)^{-1}$ (otherwise we could take $g = \mathrm{id}$) and

$$|d| \leq |c\,\mathrm{Re}(z)| + 1 \leq \frac{|\mathrm{Re}(z)| + 1}{\mathrm{Im}(z)},$$

with c, d relatively prime. We can further choose $a, b \in \mathbb{Z}$ so that $ad - bc = 1$ and gz achieves the maximum with

$$|a|, |b| \leq \max\{|c|, |d|\} \leq \frac{|\mathrm{Re}(z)| + 1}{\mathrm{Im}(z)}.$$

We next take a translation $h = \begin{pmatrix} 1 & n \\ 0 & 1 \end{pmatrix}$ such that hgz has real part between $-1/2$ and $1/2$. As shown in [484], $hgz \in F$. We estimate the height of h and then of hg. If $c \neq 0$, then $|c| \geq 1$. As $\mathrm{Im}(z) < 1$, we have $|z| \leq |\mathrm{Re}(z)| + 1$, and

$$|n| \leq |gz| + 1 \leq \frac{|az + b|}{|c||z + d/c|} + 1 \leq \frac{(|\mathrm{Re}(z)| + 1)(|\mathrm{Re}(z)| + 3)}{\mathrm{Im}(z)^2},$$

while if $c = 0$, we have $d \neq 0$ (so $|d| \geq 1$) and

$$|n| \leq |gz| + 1 \leq \frac{|az + b|}{|d|} + 1 \leq \frac{(|\mathrm{Re}(z)| + 1)(|\mathrm{Re}(z)| + 3)}{\mathrm{Im}(z)}.$$

Then

$$H(hg) = H(a + nc, b + nd, c, d) \leq \frac{(|\mathrm{Re}(z)| + 1)^2(|\mathrm{Re}(z)| + 4)}{\mathrm{Im}(z)^3},$$

which gives what we need with $\delta = 6$.   □

21.9 Proposition ([263, Lemma 5.2])   *There exists an absolute constant $c > 0$ with the following property. Let $x_1, x_2 \in Y(1)$ with $\Phi_N(x_1, x_2) = 0$ for some* integer $N \geq 1$ and $z_1, z_2 \in F$ with $j(z_i) = x_i$. Then there exists $\alpha \in \text{Mat}_2(\mathbb{Z})$ such that

$$\det \alpha = N, \quad z_2 = \alpha z_1, \quad \text{and} \quad H(\alpha) \leq cN^7$$

for some absolute positive c.

Proof It is a standard fact (see, e.g. [326, Ch. 5.2]) that there are co-prime non-negative integers a, b, d with $ad = N$ and $0 \leq b < d$ such that z_2 and $z = (az_1 + b)/d$ are in the same $\text{SL}_2(\mathbb{Z})$-orbit. Let

$$h = \begin{pmatrix} a & b \\ 0 & d \end{pmatrix},$$

so that $z = hz_1$. Since $z_1 \in F$, we have $|\text{Re}(z_1)| \leq 1/2$ and $\text{Im}(z_1) \geq 1/2$. Then $|\text{Re}(z)| \leq a|\text{Re}(z_1)|/d + b/d \leq a/(2d) + 1 \leq N/2 + 1 \leq 2N$. Next we estimate the imaginary part $\text{Im}(z) = a\text{Im}(z_1)/d \geq a/(2d) \geq 1/(2N)$. Thus, $D(z) \leq 2N$.

By Lemma 21.8 there exists $\rho \in \text{SL}_2(\mathbb{Z})$ with $\rho z \in F$ and $H(\rho) \leq cD(z) \leq c'N^6$, where $c > 0$ is the absolute constant in 21.8 and $c' = 2^6 c$. The elements $\rho z, z_2 \in F$ are in the same $\text{SL}_2(\mathbb{Z})$-orbit, hence $z_2 = \rho z = \rho h z_1$. Let $\alpha = \rho h$. Then $z_2 = \alpha z_1$, $\det \alpha = N$, and $H(\alpha) \leq 2H(\rho) \max\{a, b, d\} \leq 2c'N^7$. □

In fact, the result holds with exponent $3/2$ (see [362, Lemma 2.1]). The much more involved generalization to Shimura varieties is in [399].

Two Pairs of Linked Coordinates

The following result is essentially [263, Proposition 5.1] but proceeding under the assumption of Conjecture 21.2 rather than the assumption of asymmetry.

21.10 Proposition *Suppose the curve $V \subset Y(1)^n$ defined over a field L of finite type is not contained in any proper special subvariety, and Conjecture 21.2 holds for V. Let $i, j, k, \ell \in \{1, \ldots, n\}$ with $i \neq j, k \neq \ell$, and $\{i, j\} \neq \{k, \ell\}$. Then there are only finitely many $(x_1, \ldots, x_n) \in V$ such that there exist $M, N \geq 1$ with*

$$\Phi_M(x_i, x_j) = 0, \quad \Phi_N(x_k, x_\ell) = 0.$$

Proof We maintain the notation introduced above and suppose that V has a point $x = (x_1, \ldots, x_n)$ in its intersection with a special subvariety

$$S : \Phi_N(x_i, x_j) = 0, \quad \Phi_M(x_k, x_\ell) = 0, \quad N, M \geq 1,$$

where $\{i, j\} \neq \{k, \ell\}$ (though the sets are permitted to intersect), and we may suppose $N \geq M$. If $\{i, j\}$ and $\{k, \ell\}$ are disjoint, the complexity of S is

$\max(N, M) = N$, while if they intersect (say $j = k$), there is a further relation $\Phi_L(x_i, x_\ell) = 0$ with $1 \le L \le NM$, and the complexity could be so high as N^2. In any case, under our assumption of Conjecture 21.2, there are positive constants (depending on V) such that x has at least CN^δ conjugates over a fixed field of definition, and each such conjugate gives rise to point (α, β) on the definable set X with height at most cN^7.

However, these points may not be distinct. We first show that our CN^δ conjugates give rise to $C'N^\delta$ distinct points (of height at most cN^7) on X.

Indeed, we observe first that we may assume that $Z \cap Y_{\alpha,\beta}$ is finite for every choice of $\alpha, \beta \in \mathrm{GL}_2^+(\mathbb{R})$. For suppose, to the contrary, that some $Z \cap Y_{\alpha,\beta}$ is infinite. Being a definable set, it then contains a cell of positive dimension, and hence $j^{-1}(V) \cap Y_{\alpha,\beta}$, which is a complex analytic set, is also positive-(complex)-dimensional. However, $j^{-1}(V)$ is one-(complex)-dimensional, and we conclude that $j^{-1}(V) \subset Y_{\alpha,\beta}$. Let V' be the image of V under projection to $Y(1)^2$ on the i, j coordinates. We cannot have $\dim V' = 0$, for then the relation $\Phi_N(x_i, x_j) = 0$ holds identically on V, contrary to our hypothesis. So $\dim V' = 1$. By Modular Ax–Logarithms (Theorem 15.1), V' is contained in a proper weakly special subvariety. Since this (by hypothesis) cannot be a special subvariety, it must mean some coordinate is constant on v', and if the intersection is infinite, then the other coordinate is also constant, whence $\dim V' = 0$, which is impossible.

The set $Y_{\alpha,\beta}$ is a fibre in the definable family

$$Y = \{(z_1, \ldots, z_n, \alpha, \beta) \in \mathbb{H}^n \times \mathrm{GL}_2^+(\mathbb{R})^2 : \ z_j = \alpha z_i, \ z_\ell = \beta z_k\}.$$

The family of intersections $Z \cap Y_{\alpha,\beta}$ is definable and by Uniform Finiteness there is a uniform finite bound for the number of points in any $Y_{\alpha,\beta} \cap Z$.

Now distinct conjugates of x gives rise to distinct pre-images $z \in F^n$, hence with at least CN^δ such points, and uniformly bounded intersections with the $Y_{\alpha,\beta}$, our set X does contain at least $C'N^\delta = C''T^{\delta/7}$ rational points of height at most $T = cN^7$. Hence, by the counting theorem, if T is sufficiently large, X contains positive-dimensional semi-algebraic sets.

The idea now is the following. Our set X contains some semi-algebraic arc of (α, β) such that $Y_{\alpha,\beta}$ meets Z. If we complexify this arc we get a one-complex-dimensional family of $Y_{\alpha,\beta}$, whose union is an algebraic hypersurface $U \subset \mathbb{H}^n$, meeting X in (as we show) a set of positive real dimension and hence (as $U \cap j^{-1}(V)$ is complex analytic) a set of one complex dimension. As $\dim j^{-1}(V) = 1$, we get $j^{-1}(V)$ contained in the hypersurface U, allowing us to apply Modular Ax–Logarithms towards an eventual contradiction.

But this argument depends on the intersection $U \cap j^{-1}(V)$ being positive-dimensional, which it might not be if the intersection points $Y_{\alpha,\beta} \cap Z$ do not move as (α, β) moves along the arc. And indeed (as we have observed) X does contain positive-dimensional semi-algebraic sets that stabilize points, and in those families the intersection point would not move. So we need to work a bit more with the more refined version of the counting theorem, and with more o-minimality properties.

We apply Theorem 9.14 with $k = 1, \epsilon = \delta/14$ (say), to get a definable family $W \subset \mathbb{R}^M \times X$ of underalgebraic cells $W_y \subset X$ such that $X(\mathbb{Q}, T)$ is contained in the union of at most $cT^{\delta/14}$ fibres of W.

Let W_y be one of these fibres. Each point $(\alpha, \beta) \in W_y$ corresponds to a set $Y_{\alpha,\beta}$ that intersects Z. Those intersections are finite and of uniformly bounded cardinality. We can then find a finite number of families of definable functions

$$f_{y,i} \colon W_y \to Y_{\alpha,\beta} \cap Z$$

that parameterize all the intersection points. As the intersections are all non-empty we may assume that each $f_{y,i}$ is defined on all W_y. They are then continuously differentiable outside some lower-dimensional fibre U_y.

Let now W_y be one of the underalgebraic cells containing rational points of X. Suppose there is a point in $W_y \backslash U_y$ where the differential of some $f_{y,i}$ is non-zero. Then there is a semi-algebraic curve in W_y on which the intersection with Z is non-constant.

We now need to break into cases depending on the subset $I \subset \{i, j, k, \ell\}$ for which the corresponding coordinate function is constant on V. We can observe that I contains at least two distinct elements, one (at least) from $\{i, j\}$ and one (at least) from $\{k, \ell\}$; otherwise the existence of the point x would force V to be contained in a proper special subvariety, contrary to our hypothesis. Let $V' \subset Y(1)^{\#I}$ be the projection of V onto the coordinates in I. Then, $\dim V' = 1$.

Let us first consider the case where $I = \{i, j, k, \ell\}$, so that none of these coordinates is constant on V. The union of the $Y_{\alpha,\beta}$ over the complexification of the semi-algebraic arc (around some non-singular point) lies in a hypersurface in $\mathbb{H}^n$ that intersects Z in uncountably many points. The projection of this hypersurface on the coordinates in I is a hypersurface in $\mathbb{H}^4$ that then intersects $j^{-1}(V')$ in a set of complex dimension 1, and thus $j^{-1}(V')$ is contained in a hypersurface. By Modular Ax–Logarithms (Theorem 15.1), we conclude that $j^{-1}(V')$ is contained in a proper weakly special subvariety, but this is impossible as no coordinate is constant, so such a weakly special subvariety would in fact be special, and contradict our hypothesis.

Note that this would not lead to a contradiction if the $Y_{\alpha,\beta}$ were hypersurfaces, in which case the union could be all $\mathbb{H}^n$.

Let us consider the case $\#I = 3$. Say $k \notin I$ and so x_k constant on V. Then the union of $\{(z_i, \alpha z_i, \beta z_k)\}$ over the complexified arc is contained in a hypersurface in $\mathbb{H}^3$, and so $j^{-1}(V')$ is contained in a hypersurface, which leads (via Modular

Ax–Logarithms) to a contradiction. The more general case of $\#I = 3$ follows along the same lines. Finally, the case $\#I = 2$ is treated similarly.

Hence, each $f_{y,i}$ must have vanishing differential at all points of $W_y \setminus U_y$. The corresponding intersections with Z are constant, and account for at most

$$cT^{\delta/14}$$

points. But we have $cT^{\delta/7}$ points to account for.

Therefore, the sets U_y must account for at least $cT^{\delta/14}$ points. We repeat the application of the Counting Theorem to get underalgebraic cell families for the family U. We find, for large enough T, non-vanishing differentials give rise again to a contradiction. The dimension of the sets is decreasing. We conclude that the complexity is bounded for such points. $\square$

21.11 Remarks

1. The case with one of the coordinates among $\{i, j, k, \ell\}$ being constant on V can be handled unconditionally in a similar way to the case of one special coordinate and one linked pair considered below.
2. A different way of showing that the intersections must move on some semi-algebraic curve the idea of semi-rational points and Theorem 9.16.

One Special Coordinate and One Linked Pair

Here we do not need to assume Conjecture 21.2, and the conclusion is unconditional. This is because special points have well-controlled height. The following relies on a generalization of the Chowla–Selberg formula in [385].

21.12 Theorem ([263, Lemma 4.3])   *Let $\epsilon > 0$. There is a constant $c(\epsilon) > 0$ such that if $x \in Y(1)$ is special, then*

$$h(x) \le c(\epsilon)\Delta(x)^{\epsilon}.$$

21.13 Proposition   *Suppose the curve $V \subset Y(1)^n$ defined over a field L of finite type is not contained in any proper special subvariety. Let $i, j, k, \in \{1, \ldots, n\}$ be distinct. Then there only finitely many $(x_1, \ldots, x_n) \in V$ such that x_i is special and there exists $N \ge 1$ with*

$$\Phi_N(x_j, x_k) = 0.$$

Proof   The proof proceeds along similar lines to that of Proposition 21.10, suitably modified for the special coordinate (which leads to quadratic points in Z). The height bound Theorem 21.12 for the special coordinate controls the height of the other coordinates, via the equations for the curve. Then the Galois bounds follow via isogeny estimates; see Corollary 21.19.   □

21.14 Proof of Theorem 21.7

As we do not need to assume Conjecture 21.2 the above proves Theorem 21.7.

□

Arithmetic Conjectures

If Conjecture 21.1 holds for a curve V, then Conjecture 21.2 holds, as there are only finitely many exceptional points involved. Thus, Conjecture 21.2 is a minimal conjecture along these lines, but also a somewhat unsatisfactory one.

A more meaningful, though also much stronger, conjecture is to postulate degree growth for just likely intersections.

21.15 Conjecture   *Let $W \subset Y(1)^2$ be a curve defined over a field L of finite type. There exist positive constants $C(W), \delta(W)$ such that, for any $N \geq 1$ and any point $(x, y) \in W \cap M_N$, where M_N is the modular curve defined by $\Phi_N(x, y) = 0$,*

$$[L(x, y) : L] \geq CN^\delta.$$

Proof of Theorem 21.4   If $(x, y) \in \mathbb{C}^2$ with $\Phi_N(x, y) = 0$ then x, y are either both algebraic or both non-algebraic. On a curve that is not defined over $\overline{\mathbb{Q}}$ there can be only finitely many algebraic points. The result follows as a consequence of the following.   □

21.16 Proposition ([431, Lemma 7.3])   *Let L be a finitely generated subfield of $\mathbb{C}$. There exists positive constants c, δ (depending on L) with the following property. Let $P = (x, y) \in Y(1)^2$ be a point with non-algebraic coordinates such that $\Phi_N(x, y) = 0$. then*

$$[L(x, y) : L] \geq cN^\delta.$$

Proof   This follows from gonality properties of the modular curves; for the details see [431].   □

In fact, the following much stronger conjecture seems to be in line with expectations in conjectures for uniform boundedness of torsion in elliptic curves (see the discussion in [431, §3]).

21.17 Conjecture   *There exist positive constants c, δ with the following property. Let $x, y \in \overline{\mathbb{Q}}$, not special, with $\Phi_N(x, y) = 0$. Then*

$$[\mathbb{Q}(x, y) : \mathbb{Q}] \geq cN^\delta.$$

Clearly, this conjecture implies Conjecture 21.2 and hence Conjecture 21.1. However, weaker conjectures, with constants depending on the curve V on which the points sit, would suffice to give Conjecture 21.2, and are only needed for curves defined over $\overline{\mathbb{Q}}$. Such conjectures concern the height of the intersections of a curve $V \subset Y(1)^2$ with modular curves and are weakenings of the Bounded Height conjecture (Theorem 20.7). They imply Conjecture 21.2 via isogeny estimates ([355, 356]); the following refinement is from [235].

21.18 Theorem (Isogeny Estimate; [235, Théorème 1.4]) *Let E, E' be elliptic curves defined over a number-field K with $[K : \mathbb{Q}] \leq D$, where $D \geq 2$. If E, E' are isogenous (over $\overline{K}$), then there is an isogeny between them (over $\overline{K}$) of degree N' with*

$$N' \leq cD^2(\log D)^2 \max\left(1, h(j(E))\right)^2,$$

where c is an absolute (explicit) constant.

21.19 Corollary *Suppose that $\Phi_N(x, y) = 0$ where $x, y \in \overline{\mathbb{Q}}$ are non-special. Let $D = \max(2, [\mathbb{Q}(x, y) : \mathbb{Q}])$. Then,*

$$N \leq cD^2(\log D)^2 \max\left(1, h(x)\right)^2.$$

Proof   There is a cyclic isogeny of degree N between the elliptic curves E_x, E_y with j-invariants $x, y \in K$. Neither E_x nor E_y has CM, so N is unique. By the Isogeny Estimate, there exists an isogeny $E \to E'$ of degree N' with the given bound. However, for non-CM elliptic curves the group of homomorphisms $E \to E'$ is generated by any cyclic isogeny (see [254, Lemma 3.2]), hence $N \leq N'$.   □

One could put $\min(h(x), h(y))$ in place of $h(x)$, as only one of the heights enters the estimate, but in fact the heights of x, y related by $\Phi_N(x, y) = 0$ are close, by a result of Faltings (see, e.g. [263, p. 15]), so there is little to be gained in that direction, but one sometimes wants to consider one of x, y as fixed.

If the height $h(x)$ is suitably small in comparison to N, this implies a lower bound for D in terms of N of the required form.

We return to consider the intersection points of a given curve $V \subset Y(1)^2$ with the modular curve defined by $\Phi_N(x, y) = 0$. The analogue of the Bounded Height conjecture does not hold in the modular setting. As observed in [90], the heights of singular moduli can grow at least like the logarithm of the discriminant. A Weakly Bounded Height conjecture formulated in [254] implies the following.

21.20 Conjecture ([254, p. 47]) *Let $V \subset Y(1)^2$ be a curve defined over $\overline{\mathbb{Q}}$ and not special. There is a constant $c(V) > 0$ such that, for $(x, y) \in V$ with x, y not special and $\Phi_N(x, y) = 0$,*

$$h(x, y) \leq c(V)\log(1 + N).$$

Moreover ([254, Theorem 1.2]), such a V always has points of this form whose height grows at this rate, so the conjecture would be best possible up to the constant. Under this conjecture one would get that

$$D \gg_{V,\epsilon} N^{\frac{1}{4}-\epsilon},$$

which gives an adequate degree lower bound. It is shown in [254, Theorem 1.1] that this conjecture holds for V if $\deg_x V \neq \deg_y V$, and an elaboration of this argument in [263] shows that it holds for asymmetric curves.

21.21 Lemma ([263, Lemma 4.2]) *Let $V \subset Y(1)^n$ be an irreducible asymmetric curve defined over $\overline{\mathbb{Q}}$ that is not contained in a proper special subvariety. Then Conjecture 21.2 holds for V.*

21.22 Proof of Theorem 21.6
With the degree lower bound in Lemma 21.21 the conclusion follows from the proof of Theorem 21.3.   □

However, Conjecture 21.20 is considerably stronger than would be required to deduce Conjecture 21.2. A Super Weakly Bounded Height conjecture is formulated in [257] which would still imply Conjecture 21.2 via isogeny estimates.

21.23 Conjecture (Super-Weakly Bounded Height Conjecture; [257, Appendix B]) *In the setting of Conjecture 21.20 we have*

$$h(x, y) \leq c(V, \epsilon)N^\epsilon.$$

Even this is more than required for Conjecture 21.2, and the following would suffice.

21.24 Conjecture ("Super-Duper Weakly Bounded Height") *In the setting of 21.20 we have, for some positive η,*

$$h(x, y) \leq c(V, \eta)N^{\frac{1}{2}-\eta}.$$

The bound

$$h(x, y) \leq c(V, \epsilon)N^{1+\epsilon}$$

follows from estimates for the height of the modular polynomial ([147, 254]).

In the next chapter, we show that these same conjectural degree/height bounds would suffice to establish the full Zilber–Pink conjecture for $Y(1)^n$.

22

Conditional Modular Zilber–Pink

Here we show that the point-counting strategy, using modular Ax–Schanuel, and assuming a suitable conjecture on degrees of optimal points (LGO, Conjecture 22.1, which would follow from any of the conjectures at the end of Chapter 21), gives a full proof of ZP in the modular setting. We adapt the exposition in Habegger–Pila [264].

The first step is to show that everything can be reduced to showing finiteness of optimal points. Then the proof proceeds in a similar way to that of Theorem 21.3 but using the following in place of Conjecture 21.2.

22.1 Conjecture (Large Galois Orbit conjecture (LGO); [264, Conjecture 8.2]) *Let $V \subset Y(1)^n$ defined over some finitely generated field L. There exists positive constants $c(V), \delta(V)$ such that if $P \in V$ is an optimal point (i.e. the singleton $\{P\} \subset V$ is an optimal subvariety of V), then*

$$[L(P) : L] \geq c\Delta(\langle P \rangle)^\delta.$$

22.2 Theorem ([264, Theorem 1.2]) *Assuming Conjecture 22.1, then ZP holds for $Y(1)^n$.*

In giving the proof here we fill in a small gap left in the argument as presented in [264]. Another way to complete the argument uses [264, Corollary 7.2], as is done in the corresponding results for abelian varieties in [264]. For some unconditional special cases, see [15].

We must first introduce the *defect condition* and show that it is satisfied.

The Defect Condition

Let $V \subset Y(1)^n$, and $A \subset V$. We have already introduced the defect $\delta(A)$ and the weakly special defect $\delta_W(A)$ and the corresponding notions of optimality.

22.3 Definition ([264, Definition 4.2]) Let X be a mixed Shimura variety with its collections S, W of special and weakly special subvarieties. We say that X satisfies the *defect condition* if, whenever $A \subset B \subset X$, with A non-empty, we have

$$\delta(B) - \delta_W(B) \leq \delta(A) - \delta_W(A).$$

22.4 Proposition ([264, Proposition 4.3]) *The defect condition holds in $Y(1)^n$ and in $\mathbb{G}_m^n$, for all n.*

Proof Let $A \subset B \subset X$. First consider the case $X = \mathbb{G}_m^n$. Set

$$L(B) = \{(a_1, \ldots, a_n) \in \mathbb{Z}^n : x_1^{a_1} \cdots x_n^{a_n} \text{ is constant on } B\},$$

$$M(B) = \{(a_1, \ldots, a_n) \in \mathbb{Z}^n : x_1^{a_1} \cdots x_n^{a_n} \text{ is constant and a root of unity on } B\}.$$

We have $\mathrm{codim}\langle B \rangle_S = \mathrm{rank}(M)$ and $\mathrm{codim}\langle B \rangle_W = \mathrm{rank}(L)$. Thus,

$$\delta_S(B) - \delta_W(B) = \mathrm{rank}(L(B)/M(B))$$

is the multiplicative rank of constant monomial functions on B, and likewise for A.

However, monomial functions that are constant but multiplicatively independent on B remain so on A (which is non-empty).

Now consider $X = Y(1)^n$. Then $\delta_S(B) - \delta_W(B)$ is equal to the maximal number of constant, non-special, and Hecke-independent (i.e. in pairwise distict Hacke orbits) coordinates on $\langle B \rangle_W$, and likewose for A. Say $\{x_i, i \in I\}$ is a maximal such set for B. Since A is non-empty, so is $\langle A \rangle_W$, and since $A \subset B$ we have $\langle A \rangle_W \subset \langle B \rangle_W$. Then $\{x_i, i \in I\}$ are constant, non-special, and Hecke-independent coordinates on $\langle A \rangle_W$. □

In [264, Proposition 4.3] the defect condition is also affirmed for abelian varieties with the special subvarieties being their torsion cosets, and the following conjecture was made.

22.5 Conjecture ([264, Conjecture 4.4]) *Every mixed Shimura variety (and every weakly special subvariety) satisfies the defect condition.*

Following confirmation in the pure case in [169] and particular (product) mixed cases in [438], Conjecture 22.5 is affirmed for every mixed Shimura variety in [40] and [122].

The main point of this property for us is the following implication.

22.6 Proposition ([264, Proposition 4.5]) *Let X have the defect condition, and $V \subset X$. If $A \subset V$ is optimal for V, then A is weakly optimal for V.*

Proof Suppose $A \subset V$ and A is optimal. Consider a subvariety B with $A \subset B \subset V$ and $\delta_{\mathcal{W}}(B) \leq \delta_{\mathcal{W}}(A)$. We have

$$\delta(B) - \delta_{\mathcal{W}}(B) \leq \delta(A) - \delta_{\mathcal{W}}(A),$$

whence

$$\delta(B) = \delta_{\mathcal{W}}(B) + \delta(B) - \delta_{\mathcal{W}}(B) \leq \delta_{\mathcal{W}}(A) + \delta(A) - \delta_{\mathcal{W}}(A) = \delta(A).$$

Since A is optimal, we must have $B = A$ and so A is $\mathcal{W}$-optimal. $\square$

Reduction to Optimal Points

Let us write $\mathrm{Opt}^0(V)$ for the set of optimal points of some $V \subset Y(1)^n$. The following clearly follows from ZP.

22.7 Conjecture

1. *Let $V \subset Y(1)^n$. Then $\mathrm{Opt}^0(V)$ is a finite set.*
2. *Let $V \subset \mathbb{G}_m^n$. Then $\mathrm{Opt}^0(V)$ is a finite set.*

22.8 Proposition ([264])

1. *Suppose Conjecture 22.7.1 holds for all $V \subset Y(1)^n$, for all n. Then ZP holds for all $V \subset Y(1)^n$, for all n.*
2. *Suppose Conjecture 22.7.2 holds for all $V \subset \mathbb{G}_m^n$, for all n. Then ZP holds for all $V \subset \mathbb{G}_m^n$, for all n.*

Proof We do the proof first for $Y(1)^n$. Optimal subvarieties are weakly optimal. By o-minimality, Ax–Schanuel, and definability, the weakly optimal subvarieties of V come from finitely many families of weakly special subvarieties. Thus, elements of $\mathrm{Opt}(V)$ are components of intersections of V with translates from these finitely many families.

We now argue by induction on $\dim V$, the base case $\dim V = 0$ being trivial, and hence suppose that the conclusion holds for all subvarieties whose dimension is smaller than $\dim V$.

Fix one such family T (recall the definition of such families in Definition 4.16). Then T is the total space of the family, there is a projection $T \to Y(1)^{p_0}$ and the weakly special subvarieties in the family are the fibres of this map. Let τ be the dimension of the T_y.

It suffices to show that the number of fibres T_y for which some component of $V \cap T_y$ is optimal is finite. Let C be an optimal component of V that is a weakly optimal component and more precisely a component of $V \cap T_y$ for some y. We may assume that $T_y = \langle C \rangle_W$.

Let $V_T \subset Y(1)^{p_0}$ be the image of V under projection to $Y(1)^{p_0}$. The projection $V \to V_T$ has fibres of generic dimension outside of some proper closed subvariety $V' \subset V$. If $C \subset V'$, then it is also optimal there, and finiteness follows by induction. So we may assume that $C \not\subset V'$ and so $\dim C$ is the generic fibre dimension of $V \to V_T$.

The defect $\delta(y)$ is equal to the number of coordinates that are non-special and geodesically independent. The pre-image of $\langle \{y\} \rangle$ is then the smallest special subvariety containing C and we see that

$$\dim \langle C \rangle = \delta(y) + \tau,$$

and so

$$\delta(C) = \delta(y) + \tau - \dim C.$$

The claim now is that y is an optimal point in V_T. If not, suppose $\{y\} \subset A$, where A is a component of $V_T \cap S$ with S special and $\delta(A) \leq \delta(y)$, so that

$$\dim \langle A \rangle \leq \dim A + \delta(y).$$

Let B be the component of $V \cap \phi^{-1}(\langle A \rangle)$ containing C. Then $\dim B = \dim A + \dim C$, and $\dim \phi^{-1}(\langle A \rangle) = \dim \langle A \rangle + \tau$, and so

$$\delta(B) \leq \dim \langle A \rangle + \tau - (\dim A + \dim C) \leq \delta(y) + \tau - \dim C = \delta(C).$$

By optimality of C we have $B = C$, hence $A = \{y\}$ and $\{y\}$ is optimal.

The finiteness of such C now follows from the assumption of 21.8.

The proof in the case of $\mathbb{G}_m^n$ proceeds in a completely analogous way.   □

Finally, we show that Conjecture 22.7 is affirmed under the assumption of LGO.

22.9 Theorem ([264, Theorem 10.1])   *Assume LGO.*

Let $V \subset Y(1)^n$. Then V contains only finitely many optimal points.

Proof   Let L be a finitely generated field of definition for V. We have the uniformization $j : \mathbb{H}^n \to Y(1)^n$, with fundamental domain F. Suppose there is a special subvariety $S_0 \subset \mathbb{H}^n$ that intersects $j^{-1}(V)$ in an isolated point $Z_0 \in F$ with $j(Z_0) \subset V \cap j(S_0)$ optimal. Write $Y_0 = j(Z_0)$.

The special subvariety S_0 has a certain form corresponding to a strict partition R of the coordinates. Let us write $R = (R_0, R_1, \ldots, R_\ell)$, where R_0 is the part of variables that are constant and special, while in each of the other parts R_i, $i = 1, \ldots, \ell$, all variables are non-constant but $\mathrm{GL}_2^+(\mathbb{Q})$-related.

The family M of Möbius subvarieties of $\mathbb{H}^n$ corresponding to the partition R is definable, parameterized by $P = \mathbb{H}^{R_0} \times \mathrm{GL}_2^+(\mathbb{R})^{\Sigma(\#R_i - 1)}$, the fibre over the parameter $t \in P$ being denoted M_t. A special subvariety corresponds to a parameter t that has rational $\mathrm{GL}_2^+(\mathbb{R})$ coordinates (though may also be given by non-rational choices), and quadratic $\mathbb{H}$ coordinates. We call such parameters rational points.

We have the definable subset

$$Z = \{(t,u) : t \in P, u \in M_t \cap j^{-1}(V) \cap F\}.$$

The conjugates of Y_0 over L are also optimal points of V. Under LGO, Y_0 has many conjugates and they give rise to many points $(t,u) \in Z$ with rational t. By the Counting Theorem, if $\Delta(j(S_0))$ is sufficiently large, then Z has points (t,u), where t lies in positive-dimensional semi-algebraic sets. Arguing as in the proof of Theorem 21.3, there must be a one-parameter real semi-algebraic family of Möbius subvarieties $S_t = M_{p(t)}$, of form R, that intersect $j^{-1}(V)$, and whose intersection with $j^{-1}(V)$ is not a constant point but moves with the parameter.

The condition that all the S_t have the form R means that for each variable z_i, $i \in R_0$, we have some algebraic function

$$c_i : (1,1) \to \mathbb{H},$$

with $c_i(0) = (Z_0)_i$ and special, and for each $i,j \in R_k$, $k = 1, \ldots, \ell$, we have an algebraic function

$$g_{ij} : (-1,1) \to \mathrm{GL}_2^+(\mathbb{R})$$

such that the Möbius variety S_t is described on the corresponding coordinates by the equations

$$z_i = g_{ij}(t)z_j, \quad g_{ij}(0) \in \mathrm{GL}_2^+(\mathbb{Q}), \quad (Z_0)_i = g_{ij}(0)(Z_0)_j.$$

By o-minimality, and since we have many conjugates, and the algebraic functions have some fixed degree (depending on some choice of an ϵ in the Counting Theorem), we can assume that all the algebraic functions are non-singular around $t = 0$.

We now complexify the parameter t around $t = 0$ and get a one complex-parameter algebraic family of (complex) Möbius subvarieties (as matrix entries may now be complex) S_t. The Zariski closure S of the union of the S_t intersects $j^{-1}(V)$ in a one-complex-dimensional component A, and we have

$$\dim S - \dim A = \dim S_0.$$

By Modular Ax–Schanuel, we have some component B with $A \subset B$ whose Zariski closure $\overline{B}$ is weakly special and satisfies

$$\dim \overline{B} - \dim B \leq \dim \overline{A} - \dim A,$$

where $\overline{A} \subset S$ is the Zariski closure of A.

Let $\langle B \rangle$ be the smallest special subvariety containing B. We want to show that

$$\dim \langle B \rangle - \dim B \leq \dim S - \dim A = \dim S_0 - \dim\{Z_0\} = \dim S_0,$$

which would imply that $\{y_0\}$ was not optimal after all, and the contradiction implies that the complexity is bounded, concluding the proof.

If $\overline{A} = S$, then $S_0 \subset S = \overline{A} \subset \overline{B}$ is special, so that $\langle B \rangle = \overline{B}$ and what we need follows from the conclusion afforded by Ax–Schanuel.

So it remains to consider what happens if $\overline{A}$ is smaller than S. The claim is that it suffices to show that

$$\dim \langle B \rangle - \dim \overline{B} \leq \dim S - \dim \overline{A}.$$

For then

$$\begin{aligned}
\dim \langle B \rangle - \dim B &\leq \dim S - \dim \overline{A} + \dim \overline{B} - \dim B \\
&= \dim S - \dim A + (\dim \overline{B} - \dim B) - (\dim \overline{A} - \dim A) \\
&\leq \dim S - \dim A
\end{aligned}$$

as required. The proof is thus concluded with the following proposition.

22.10 Proposition *With the notation as above we indeed have*

$$\dim \langle B \rangle - \dim \overline{B} \leq \dim S - \dim \overline{A}.$$

Proof Since $\overline{B}$ is weakly special,

$$k = \dim \langle B \rangle - \dim \overline{B}$$

is equal to the maximal number of coordinate functions that are constant, non-special, and $\mathrm{GL}_2^+(\mathbb{Q})$-independent on $\overline{B}$. These coordinates are then constant, non-special, and $\mathrm{GL}_2^+(\mathbb{Q})$-independent on $\overline{A}$.

Such a coordinate cannot be in R_0, because its constant value would need to equal its value at $t = 0$, which is special. And no two such coordinates can be in the same R_i, $i \geq 1$, for then at $t = 0$ they would be $\mathrm{GL}_2^+(\mathbb{Q})$ related. So these k coordinates come from different R_i, $i \geq 1$ (and so $k \leq \ell$).

Consider now the subvariety $S^* \subset S$ in which we fix each such coordinate to equal its constant value. Then

$$\dim S^* = \dim S - k.$$

This can be seen fibre-wise: we have reduced the dimension of each fibre by k. Since these coordinates are also constant on A, we have $A \subset S^*$ and hence $\overline{A} \subset S^*$. This gives what we need. $\qquad\qquad\square$

This completes the proof of the theorem. $\qquad\qquad\square$

22.11 Remark One can remove B from the argument noting that, by the defect condition,

$$\dim\langle B\rangle - \dim\langle B\rangle_W \leq \dim\langle A\rangle - \dim\langle A\rangle_W.$$

Now the proof proceeds in the same way. Essentially, one is showing that S can be replaced with the S^* coming from a family of lower-dimensional Möbius subvarieties.

23

O-Minimal Uniformity

In this chapter and the next we describe two kinds of uniformity. The first is the uniformity that comes with o-minimality, considered in this chapter. In the next chapter we explore a different kind of uniformity, showing, in the basic settings $\mathbb{G}_m^n$ and $Y(1)^n$, that ZP implies a uniform version of itself. This uniformity is inherent in ZP itself, rather than being an artefact of the proof method.

Results in o-minimality generally come with uniformity in definable families. This is essentially due to the cell decomposition theorem. This uniformity was crucial in the proof of the Counting Theorem (uniformity in the number of maps in an r-parameterization of the fibres in a definable family).

23.1 Definition   For $V \subset Y(1)^n$ define the *special set* of V, denoted $\mathrm{Spc}(V)$, to be the union of special subvarieties $T \subset V$.

Thus, AO is the assertion that $\mathrm{Spc}(V)$ is a finite union. An elaboration of the following result is in [427, §13].

23.2 Theorem   *Let n, d, k be positive integers, and consider the family $V \subset Y(1)^n \times P$ of hypersurfaces of degree d. There is only a finite set of possibilities for $\mathrm{Spc}(V_p)$ when V_p is defined over a number-field of degree at most k over $\mathbb{Q}$.*

Clearly, such a statement cannot hold without some restriction on the field of definition. Note that the known effective results ([317, Theorem 3; 318]) do not quite yield this uniformity. If one has further information about the possible positive-dimensional weakly special subvarieties of the V_p, then one can state more precise results.

For example, consider the family of linear subvarieties of some $Y(1)^n$. It is shown in [64, 2.3] or [429, 7.1] that a special subvariety of a linear subvariety of $Y(1)^n$ is linear and defined by equations of the form $x_i = x_j$ for some number of pairs of distinct coordinates $i, j \in \{1, \ldots, n\}$, and $x_k = \sigma_k$, where σ_k is special, for some number of coordinates k. For such a linear special subvariety

Σ call a coordinate ℓ *free* if it is not governed by any equation. Consider a linear hyperplane $L \subset Y(1)^n$ defined by

$$L : a_1 x_1 + \cdots + a_n x_n = 0, \quad a_1 \ldots a_n \neq 0.$$

If $\Sigma \subset L$ is a special subvariety, and hence linear, it can have no free coordinates. With this one can establish a modular analogue of Mann's Theorem [342] on linear dependencies among roots of unity.

23.3 Definition   Let $K \subset \mathbb{C}$ be a field. An n-tuple $(c_1, \ldots, c_n) \in \mathbb{C}^n$ is called *minimally linearly dependent* over K if $c_1, \ldots, c_n$ are linearly dependent over K but no proper subset of them is.

23.4 Theorem ([429, 1.3])   *Let $n, k \geq 1$. There exist only finitely many n-tuples of distinct non-zero singular moduli that are minimally linearly dependent over a field K with $[K : \mathbb{Q}] \leq k$.*

The conclusion of course fails for complex linear subvarieties, as any special point lies on such a subvariety. However, a numerical uniformity holds by "automatic uniformity" ([467]): Given n there is a uniform bound on the number of n-tuples of distinct singular moduli that may lie on any proper linear hypersurface. (This is the kind of uniformity addressed in Chapter 24.) Moreover, the result of [64] provides, given a linear variety over $\overline{\mathbb{Q}}$, an effective bound on the discriminants of its isolated special points.

In the following result, point-counting is applied to a set that depends on a solution to the diophantine problem, but the sets concerned lie in a definable family.

23.5 Theorem ([431, Theorem 5.2])   *Let $V \subset Y(1)^2$ be a curve that is not weakly special.*

1. *Assume Conjecture 21.17. Then there exist only finitely many (N, M) for which there exist non-special rational numbers x, y and points $(u, v) \in V$ such that $\Phi_N(x, u) = 0, \Phi_M(y, v) = 0$.*
2. *Unconditionally, there exist only finitely many such N, M for which $\max(N, M)$ is a prime number.*

23.6 Remarks

1. If V is weakly special, then the conclusion can fail.
2. The theorem excludes special x, y as in that case u, v would also be special, and though there can only be finitely many such points, there can then be infinitely many (N, M).
3. Theorem 23.5 admits a similar uniformity to Theorem 23.2.

Proof Suppose first that V is not defined over $\overline{\mathbb{Q}}$. Then it has only finitely many points defined over $\overline{\mathbb{Q}}$, and for each such point (u, v) if we seek rational x, y with the required property the isogeny estimates directly imply that N, M are bounded. So we may suppose that V is defined over $\overline{\mathbb{Q}}$.

We consider the $\mathbb{R}_{\text{an exp}}$-definable family of sets in $GL_2^+(\mathbb{R})^2$, parameterized by $Q = (z, w) \in \mathbb{H}^2$,

$$Z_Q = \{(g, h) \in GL_2^+(\mathbb{R})^2 : gz, hw \in F \text{ and } (j(gz), j(hw)) \in V\}.$$

Suppose that $(x, y) \in \mathbb{Q}^2$ and $(u, v) \in V$ exist with the required properties for some (N, M). Put $L = \max(N, M)$. By Conjecture 21.17 we have

$$[\mathbb{Q}(u, v) : \mathbb{Q}] \geq cL^{\delta}.$$

Thus, we have at least that many conjugate points $(u', v') \in V$ over $\mathbb{Q}$ and each satisfies

$$\Phi_N(x, u') = 0, \quad \Phi_M(y, v') = 0.$$

Take $z, w \in F$ with $j(z) = x, j(w) = y$ and put $Q = (z, w)$. Each pair (u', v') gives rise to a rational point on Z_Q of height bounded (by Proposition 21.9) by

$$c'L^7.$$

The Counting Theorem is uniform over the family Z_Q. So if L is sufficiently large, then Z_Q contains some positive-dimensional semi-algebraic curve. The corresponding points (gz, hw) however cannot be constant as the semi-algebraic curves in Z_Q need to account for many different (u', v'). So we get a real algebraic curve contained in

$$\{(z, w) \in \mathbb{H}^2 : (j(z), j(w)) \in V\}.$$

But then this set itself is complex algebraic, and V is bi-algebraic, hence weakly special. This contradiction proves that L is bounded.

This proves the conditional statement. If $L = \max(N, M)$ is a prime number, then a Galois bound of the required form holds, by results of [386]; see also [431]. This gives the second statement. $\qquad\square$

Note that, for fixed N, M, the existence of such points amounts to existence of rational points (x, y) on suitable curves obtained by elimination of u, v.

A similar uniformity is used by Masser–Zannier [362] in proving the existence of Hodge generic abelian varieties in dimension $g \geq 4$, defined over $\overline{\mathbb{Q}}$, and isogenous to no Jacobian variety. The existence of an abelian variety defined over $\overline{\mathbb{Q}}$ and isogenous to no Jacobian variety was proved, assuming the then conjectural AO for $\mathcal{A}_g$, in [124]. The dependence on AO was removed in

[503] but the examples produced in both cases have CM. The treatment in [362] gives examples without CM (answering a question of [124]) and moreover with the stronger property of Hodge-genericity. And the examples are defined over fields of low degree (again in contrast to CM examples).

23.7 Theorem ([362])   *Given a hypersurface $V \subset \mathcal{A}_g$, where $g \geq 2$, there exists a Hodge generic $x \in \mathcal{A}_g$ defined over a number-field of degree at most 2^{16g^2} such that A_x is not isogenous to A_y for any $y \in V$.*

We prove a toy version of this to illustrate the role of uniformity in point-counting. The argument also relies on the weak dependence of isogeny estimates on heights. We use the following definition though note that the collection of abelian varieties that are isogenous to a given one is wider than the Hecke orbit (or even the generalized Hecke orbit; see Conjecture 18.11).

23.8 Definition   Let $V \subset Y(1)^n$. We say that $(x, y) \in Y(1)^2$ *avoids* V if $\Sigma(x, y) \cap V = \emptyset$.

23.9 Theorem (After [362])   *Let $V \subset Y(1)^2$ be a curve. Then there exists $(x, y) \in Y(1)^2(\mathbb{Q})$, which avoids V.*

Proof   We show that we can assume that V is not weakly special. For suppose, first, that V is weakly special but not special, say with $x = c$ constant and not special. We can evidently assume $c \in \overline{\mathbb{Q}}$, and we are faced with the one-dimensional version of the same problem: find $x \in Y(1)(\mathbb{Q})$, which avoids c; equivalently, $x \notin \Sigma(c)$.

Now we apply isogeny estimates. If x is in $\Sigma(c)$, then there is an isogeny of degree $N \ll h(x)^2$. On the other hand, given N, the degree of Φ_N is given by the Dedekind ψ function $\psi(N)$, and satisfies

$$\psi(N) \leq \sigma(N) = \sum_{d \mid N} d \ll CN \log \log N.$$

So if we consider rational numbers up to (multiplicative) height T, of which there are $\asymp T^2$, then at most $(\log T)^5$ of them can be $\Sigma(c)$. This argument also settles the case that V is special but not strongly special, as only a finite number of rational numbers (13 in fact) are special.

Now suppose V is strongly special. Then $u \in \Sigma(v)$ for every $(u, v) \in V$, so if (x, y) is in $\Sigma(u, v)$, where $(u, v) \in V$, then $x \in \Sigma(y)$. If we consider again rational x, y up to height T, of which there are $\asymp T^4$ pairs, each x has at most $(\log T)^5$ such y in its Hecke orbit. So there are nearly T^2 pairs (x_i, y_i) in which the x_i, y_i are pairwise Hecke inequivalent.

Finally, we consider the case in which V is not weakly special. If V is not defined over $\overline{\mathbb{Q}}$, then it has only finitely many algebraic points, and we can avoid these by the same arguments as above. So we can suppose $V/\overline{\mathbb{Q}}$.

If $(u, v) \in V \cap \Sigma(x, y)$, then, for some N, M, we have

$$\Phi_N(x, u) = 0, \quad \Phi_M(y, v) = 0, \quad F(u, v) = 0,$$

where $F(u, v) = 0$ defines V, say of degree d. Eliminating u, v leads to a non-trivial algebraic relation $G_{N,M}(x, y) = 0$ between x, y of degree at most $d\psi(N)\psi(M)$.

Let us now consider $x, y \in \mathbb{Q}$ up to height T and restrict N, M to be at most $(\log T)^4$. The number of such (x, y) in such a relation is then at most

$$c(V)T^2(\log T)^{10},$$

compared with $\asymp T^4$ rational pairs (x, y) up to height T. Only a finite number of rational numbers are special so for all $T \geq c_0(V)$, where $c_0(V) \geq e$, one can find non-special $x, y \in \mathbb{Q}$ such that (x, y) has no small isogeny into V.

Fix T and suppose $x, y \in Y(1)(\mathbb{Q})$ are non-special, with $H(x), H(y) \leq T$, and (x, y) has no small isogeny to V; that is, $N, M \leq (\log T)^4$. Suppose that we have $(u, v) \in \Sigma(x) \cap V$, where $\Phi_N(x, u) = 0, \Phi_M(y, v) = 0$ with, say $N \geq (\log T)^4$ (the argument is the same if it is $M \geq (\log T)^4$). Since $\max(1, h(x)) \leq \log T$, the isogeny estimate gives

$$N \leq cD^2(\log D)^2(\log T)^2,$$

where $D = [\mathbb{Q}(u) : \mathbb{Q}]$ and we find

$$D \gg N^{1/5}.$$

Thus, (u, v) has many conjugates (over a field of definition of V). Using the same definable sets as in Theorem 23.5, if N is sufficiently large this leads to the contradictory conclusion that V is weakly special. Thus, for sufficiently large T, the Hecke orbit of (x, y) cannot intersect V. $\square$

If one considers atypical intersections rather than special points, then o-minimality does not immediately give the same kind of uniformity as in Theorem 23.2 the subvariety and the degree lower bound might depend on the particular variety. However, under the strongest arithmetic conjectures some uniformity is restored.

23.10 Proposition *Assume Conjecture 21.17. Let d, k be given and consider the family $V \subset Y(1)^3 \times P$ of curves of degree d. Then there is a uniform bound* *on the size of* Opt(V_p) *when V_p is defined over a number-field of degree at most k over* $\mathbb{Q}$, *and not contained in a proper special subvariety of* $Y(1)^3$.

Proof   Suppose V_p is not contained in a proper special subvariety, but is defined over a number-field K with $[K : \mathbb{Q}] \leq k$. Then Opt(V_p) is a set of points defined over $\overline{\mathbb{Q}}$. By Conjecture 21.17, the degree over $\mathbb{Q}$ of each point $Q \in$ Opt(V_p) is bounded below by a positive power of its complexity, independent of V_p. As the varieties are in a definable family, the point-counting applies uniformly, and one finds that, given k, the complexity is bounded by some $B(d, k)$.

Now Q is either of the form (x, y, z), where x, y, z are non-special and

$$\Phi_N(x, y) = 0, \quad \Phi_M(y, z) = 0.$$

For such points we see that $\max(N, M)$ is bounded, and so the number of points in the intersection with V_p, which has degree d, is uniformly bounded.

Or Q may have the form (up to permuting the variables) $Q = (x, y, z)$, where x is special and $\Phi_N(y, z) = 0$, where now N and $\Delta(x)$ are bounded. Then x is restricted to a finite set, and the intersections of the projection of V to the yz-plane with $\Phi_N(y, z) = 0$ has bounded size.

Lastly, Q could be of the form (x, y, z), where x, y are special. Now the conclusion follows from uniformity in AO for the projection to the xy-plane.   □

24

Uniform Zilber–Pink

From ZP to Uniform ZP

We consider ZP in the modular and multiplicative settings. For each of these settings we show that ZP implies a uniform version (UZP) of itself, in the sense of Scanlon [467] (see Conjecture 24.1). Otherwise put, in an algebraic family of subvarieties $V_p \subset X \times P$, where $X = Y(1)^n$ or $X = \mathbb{G}_m^n$, the optimal cycles $\mathrm{Opt}(V_p), p \in P$ are bounded as cycles.

For the corresponding special point problems (multiplicative Manin–Mumford and André–Oort; and also for Mordell–Lang) in a fixed ambient variety, this fact has been shown by Scanlon [467]. An argument that multiplicative ZP implies UZP is sketched by Zannier [544], illustrated in the case of the theorem of [87] concerning a curve in $\mathbb{G}_m^n$, and [352] carries out such a proof in the case of lines.

The question of uniform bounds on the intersection of a curve embedded in its Jacobian with a finite rank subgroup was raised by Mazur [367]. Could the size of the intersection be bounded depending only on the genus g of the curve and the rank r of the group? Stoll [496] shows that Pink's conjecture on families of abelian varieties implies an affirmative answer. The paper [496] also proves some uniformities (for hyperelliptic curves with $r \leq g - 3$, getting a bound depending also on the field of definition). More general results along the same lines are proved in [302] (bounds depending on g, r, and the degree d of the defining field). A bound of the form $c(d, g)^r$ is obtained in [182] and a full Uniform Mordell–Lang conjecture is established in [231]. For a uniform bound on torsion points for a family of curves over $\mathbb{C}$, see [175], generalized in [321]. For an effective bound on the height of rational points under additional conditions, see [132].

Here we use the reduction of ZP to finiteness of optimal points (given in the paper [264] and used in Chapter 21). We then carry out an argument along

205

the lines sketched by Zannier to get the uniformity implication. Some care is required to deal with the families of weakly special subvarieties that intervene in the argument. Another approach to uniformity in parameters is given by Zilber [550, Theorem 1].

Let X be a mixed Shimura variety and $V \subset X$. Under ZP, the set of optimal subvarieties of V is finite. Recall that we have defined the *optimal cycle* $\mathrm{Opt}(V_p)$ of V to be the formal sum of optimal subvarieties.

24.1 Conjecture (Uniform Zilber–Pink, UZP) *Let X be a mixed Shimura variety, and $V \subset X \times P$ a family of subvarieties of X parameterized by the constructible set P. Then there is a family of closed algebraic subsets $W \subset X \times Q$ such that, for every $p \in P$, the optimal cycle $\mathrm{Opt}(V_p)$ of V_p appears as a fibre of W. That is, for all $p \in P \exists q \in Q : \mathrm{Opt}(V_p) = W_q$.*

24.2 Theorem *Let $X = Y(1)$ or $\mathbb{G}_m$ and assume that ZP holds for X^n for all n. Then Optimal UZP holds for X^n for all n.*

This form of uniformity in special-point problems is precisely the automatic uniformity established in [467]: over an algebraic family of subvarieties, the special set of the fibres lie in an algebraic family. It would be desirable to extend this implication to the general case; that is, to show, for a Shimura variety X, that the truth of ZP for all powers X^n implies UZP for X.

Reduction to Optimal Points

We first show that the uniformity can be reduced to the case of points: it suffices to show that in a family $V \subset X \times P$ of subvarieties $V_p \subset X$, the number of optimal points in a fibre is uniformly bounded. This is what we mean by the statement that "optimal points are uniformly bounded in families". The following is then just a definable family version of Proposition 22.8.

24.3 Proposition *Let $X = Y(1)^n$ or $X = \mathbb{G}_m^n$. Assume that $\mathrm{Opt}^0(V)$ is finite uniformly in families $V \subset X$. Then UZP holds for X.*

Proof We assume that optimal points are uniformly bounded in families and show that UZP holds. Let $V \subset X^n \times P$ be a family of subvarieties.

Optimal subvarieties are geodesic optimal by Proposition 22.6. By o-minimality, definability, and Ax–Schanuel the weakly optimal subvarieties of V_p, for all p, come from finitely many such families. Thus, elements of $\mathrm{Opt}(V_p)$ are intersections of V_p with translates from these finitely many families.

Fix one such family $\phi : T \to Y$ as in Definition 3.12, in the multiplicative case, or Definition 4.15.3 in the modular case, with weakly special fibres

$T_y = \phi^{-1}(y), y \in Y$. It suffices to show that the number of translates T_y for which some component of $V_p \cap T_y$ is optimal is finite, uniformly over $p \in P$. Let $\tau = \dim T_y$.

Let $W = V \cap T$ be the family of intersections of fibres of V with T. Thus, for any p, y, $W_p \cap T_y = V_p \cap T_y$. The images $\phi(W_p) = Z_p$ are the fibres of a family $Z \subset Y \times P$ of subvarieties of Y, parameterized by P.

Since we are in characteristic zero, there is a Zariski open (in each fibre W_p) subvariety W'_p on which the restriction of ϕ is smooth of relative dimension v, with image $\phi(W'_p)$ Zariski open in its Zariski closure Z_p. Suppose that $A \subset V_p \cap T_y$ is an optimal subvariety of V. If $A \cap W'_p = \emptyset$, then $A \subset W_y \backslash W'_y$ and is optimal for the component it is in. Since $\dim W'_p < \dim W_p$, we get finiteness of such optimal A by induction on fibre dimension of the families we are considering.

So we can assume that $A \cap W'_p \neq \emptyset$. Then $A \cap W'_p$ is an irreducible component of a fibre of $\pi|_{W'_p}$ and so $\dim A = v$, and $\pi(A) = y$. The claim is that $\{y\}$ is an optimal point of Z_p.

We can assume that we have already dealt with any family of smaller weakly special subvarieties that might give rise to A. Hence, we can assume that $T_y = \langle A \rangle_W$. Then, $\dim \langle A \rangle = \dim T_y + \dim \langle \{y\} \rangle$. Thus,

$$\delta(A) = \dim \langle \{y\} \rangle + \tau - v.$$

Suppose $\{y\} \subset B, \{y\} \neq B$ with $\delta(B) \leq \delta(\{y\}) = \dim \langle \{y\} \rangle$. Let C be the component of the pre-image of B in W'_p containing A. Then,

$$\delta(C) = \dim \langle C \rangle - \dim C = \dim \langle B \rangle + \tau - (\dim B + v) \leq \dim \langle \{y\} \rangle + \tau - v = \delta(A).$$

Then $C = A$ by optimality of A and so $B = \{y\}$, and $\{y\}$ is optimal.

Now $\mathrm{Opt}^0(Z_p)$ is uniformly finite by hypothesis, and so only finitely many optimal $A \subset V_p$ come from T, uniformly over P. □

Incidence Varieties

We consider a family $V \subset X \times P$, where P is some constructible parameter space, of subvarieties $V_p \subset X$ (as usual, subvarieties are assumed to be relatively closed). The *fibre dimension* of the family is the maximum dimension of a fibre.

Associated with a family V and a positive integer h is the incidence variety

$$\mathrm{Inc}^h(V) = \{(z_1, \ldots, z_h) \in X^h : \exists p \in P : z_i \in V_p, i = 1, \ldots, h\}.$$

Since P is only assumed to be constructible, the incidence variety may not be closed (in X^h). We denote by $V^{\langle h \rangle}$ its Zariski closure (in X^h), in which $\mathrm{Inc}^h(V)$ is Zariski-dense. In particular, $V^{\langle 1 \rangle} \subset X$ is the Zariski closure of the union of all the fibres.

Uniform Finiteness of Optimal Points: Modular Case

The proof that ZP implies UZP goes via an induction on families of weakly special subvarieties. So the first step is to set up a finer description, with notation, for such families.

Families of Modular Weakly Special Subvarieties

We must introduce a finer notion of families of weakly special subvarieties than the one given in Definition 4.15. The description there uses the constant coordinates to parameterize weakly special subvarieties, and each weakly special subvariety is contained in such a family. But now we want to consider an arbitrary special subvariety as the ambient and put its weakly special subvarieties into families.

So now we formulate a slightly different definition than the one in 4.15, in which consider a special subvariety $T = T[p, g] \subset Y(1)^n$ as being determined by a partition $p = (p_1, \ldots, p_k)$ of $\{1, \ldots, n\}$ in the usual sense with no designated "p_0" for the constant coordinates, and suitable group data g, as in Definition 4.14 relating coordinates in the same p_i. If we now choose a subset $U \subset \{1, \ldots, k\}$ and let $q = (p_i, i \in U)$, then q is a partition of some subset $N \subset \{1, \ldots, n\}$, the natural projection map

$$Y(1)^n \to Y(1)^N$$

maps $T[p, g]$ onto some special subvariety

$$S(U) = S[p, g, U] \subset Y(1)^N.$$

We take

$$T[p, g, U] \subset T[p, g] \times S[p, g, U]$$

to be the graph of the projection, considered as the family of its fibres. The fibres $T[p, g, U]_t, t \in S[p, g, U]$ are then a family of weakly special subvarieties of $T[p, g]$ determined by the choice of U.

We denote the family by $T[U] = T[p, g, U]$, generally suppressing mention of p, g, and its fibres as $T[U]_t$, where $t \in S[U] \subset Y(1)^N$. We have $\dim T[p, g] = k$ and the dimension of the fibres of $T[U]$ is

$$\dim T[U]_t = k - \#U.$$

We introduce a partial order on families of weakly special subvarieties of $Y(1)^n$ as follows. If

$$T_1 = T[p_1, g_1], \quad T_2 = T[p_2, g_2]$$

are special subvarieties of $Y(1)^n$, then we say

$$T_1[U_1] < T_2[U_2]$$

if either T_1 is a proper special subvariety of T_2 or if $T_1 = T_2$ and U_1 is a proper superset of U_2. In this latter case the fibres of $T_1[U_1]$ are of lower dimension than those of $T_1[U_2]$.

24.4 Theorem *Assume ZP holds for $X = Y(1)^n$ for all n. Then $\mathrm{Opt}^0(V)$ is finite uniformly in families $V \subset X$.*

Proof We consider a family $V \subset X \times P$ where $X = Y(1)^n$. The proof is by induction on (firstly) dim P. The case dim $P = 0$ is immediate by the assumption of ZP. We may then assume that the Zariski closure of P is irreducible and the fibre dimension v of V is constant.

By o-minimality the smallest weakly special subvarieties containing the fibres $V_p, p \in P$ come in finitely many families. By induction we may then assume that there is a single family $T[U] = T[p, g, U]$ of weakly special subvarieties consisting of the fibres of

$$\phi : T \to S[U],$$

such that for all $p \in P$ there exists $t \in S[U]$ such that $\langle V_p \rangle_{\mathrm{ws}} = \phi^{-1}(t) = T[U]_t$.

By induction, we may assume the result for all families where the fibre dimension v is smaller (the dimension zero case is trivially uniform), for all families V' where the ambient family of weakly special subvarieties T' has $T' < T$ (the base cases, where dim $T = 0$ or even dim $T = 1$, are trivial).

We choose a suitable positive integer h (to be fixed below) and consider the subvariety $V^{\langle h \rangle} \subset X^h$. Then,

$$V^{\langle h \rangle} \subset T[U]^{\langle h \rangle} \subset T^h \subset X^h.$$

Further, $T[U]^{\langle h \rangle}$ is special. More precisely, if the coordinate with index $k \in \{1, \ldots, n\}$ is constant on every member of the family $T[U]$, then $z \in T[U]^{\langle h \rangle}$ satisfies $z_{ik} = z_{jk}$ for all i, j. Here we view $z_{ik}, k = 1, \ldots, n$ as a coordinate system on the ith factor in the product X^h. Then, $T[U]^{\langle h \rangle} \subset X^h$ is the special subvariety defined by such equations with k ranging over a set consisting of one element of each $p_i, i \in U$. Thus,

$$\mathrm{codim}\big(T[U]^{\langle h \rangle}, T^h\big) = \#U \cdot (h - 1).$$

Now suppose $y \in P$ and that $z_1, \ldots, z_h \in V_y$ are optimal points. Then they are atypical as point subvarieties (unless dim $V_y = 0$, in which case our finiteness assertion is trivial) for V_y in $\langle V_y \rangle$ and so there is a special subvariety $Z_i \subset T$ with $z_i \in Z_i$ and

$$\dim Z_i + v < \dim T.$$

Let

$$Z = Z_1 \times \cdots \times Z_h.$$

Since the equations defining $T[U]^{\langle h \rangle}$ are all *between* different groups of variables, while those defining the Z_i are *within* groups, the codimensions add and we have

$$\dim Z \cap T[U]^{\langle h \rangle} = \dim Z - \operatorname{codim}\bigl(T[U]^{\langle h \rangle}, T^h\bigr).$$

Then $z = (z_1, \ldots, z_h) \in Z \cap T[U]^{\langle h \rangle}$, which is a special subvariety of X^h, and is atypical for $V^{\langle h \rangle}$ as a subvariety of $T^{\langle h \rangle}$ provided that

$$\dim V^{\langle h \rangle} + \dim Z \cap T[U]^{\langle h \rangle} < \dim T[U]^{\langle h \rangle}$$

and since $\dim Z_i \le \dim X - v - 1$, we find that z is atypical provided that

$$\dim P + hv + h(\dim T - v - 1) - \operatorname{codim}\bigl(T[U]^{\langle h \rangle}, T^h\bigr) < \dim T[U]^{\langle h \rangle},$$

where

$$\dim T[U]^{\langle h \rangle} = h \dim T - \operatorname{codim}\bigl(T[U]^{\langle h \rangle}, T^h\bigr).$$

So such z is atypical provided that $h > \dim P$, which we henceforth assume.

Then by ZP, such points z are contained in one of finitely many proper special subvarieties of $Y_i \subset T^{\langle h \rangle}$. Therefore, there is a finite list of special relations such that all such z satisfy (at least) one of them. These have the following four possible forms:

(i) $z_{ij} = \sigma$ for some singular modulus σ;
(ii) $\Phi_N(z_{ij}, z_{ik}) = 0$ for some choice of i, j, k with $j \ne k$, and N;
(iii) $\Phi_M(z_{ik}, z_{jk}) = 0$ for some choice of i, j, k with $i \ne j$, and M;
(iv) $\Phi_L(z_{ij}, z_{k\ell}) = 0$ for some choice of i, j, k, ℓ, all distinct, and L.

Since these relations give proper special subvarieties of $T[U]^{\langle h \rangle}$, we note that those of type (i) and (ii) are not relations that hold identically for fibres of V. For the same reason, if i, j in a relation of type (iii) are coordinates that are constant on fibres of V, then we have $M \ne 1$.

Suppose that there are distinct optimal points z_i, $i = 1, \ldots, H$ on some fibre of V. If we select h of them and consider them in the index order, then the resulting point of X^h satisfies one of finitely many relations as above.

By the Hypergraph Ramsey Theorem (see, e.g. [247]), if we have H' large enough, then we can find a subset of H of them such that all subsets of h, taken in order, satisfy the same relation. If H is large enough (depending on the M, L occurring in the new relations), we get many points on a fibre of a family that have either a smaller enveloping special subvariety (in case of relations of type (i) or (ii) or (iii) with i, j constant coordinates) or additional constant coordinates (in other cases of (iii) or (iv)).

Only finitely many such families arise, and the optimal points remain optimal on the (possibly) smaller fibres. By induction we may assume that the number of optimal points on a fibre of any of these is uniformly bounded.

So we get a uniform bound for the number of optimal points on a fibre of V, and conclude that UZP holds. □

24.5 Remarks

1. Observe that if the family V is defined over $\overline{\mathbb{Q}}$, then W is defined over $\overline{\mathbb{Q}}$. Moreover, every variety $V_0 \subset \mathbb{C}^n$ is a fibre of some family of varieties defined over $\overline{\mathbb{Q}}$. Hence, ZP for varieties defined over $\overline{\mathbb{Q}}$ implies ZP for varieties defined over $\mathbb{C}$. In the multiplicative setting this has been shown in a much more precise form, in [92]; in the above form it follows in the multiplicative setting from Theorem 24.7.

2. Note further that the argument above is effective. If one assumes that ZP holds effectively, for varieties defined over $\overline{\mathbb{Q}}$, then it holds effectively and uniformly for families defined over $\mathbb{C}$ (where varieties over $\mathbb{C}$ are viewed as members of a suitable family defined over $\overline{\mathbb{Q}}$). This also holds for the multiplicative version below.

24.6 Proof of Theorem 24.2 in the Modular Case

This follows in view of Proposition 24.3 and Theorem 24.4. □

Uniform Finiteness of Optimal Points: Multiplicative Case

This is similar, and we set up a parallel notation for families of weakly special subvarieties. However, the proof plays out in a slightly different way due to the different nature of the special relations. This suggests that the general case, and especially mixed settings, could entail some interesting properties of families of weakly special subvarieties.

Families of Multiplicative Weakly Special Subvarieties

We introduce an analogous description in the multiplicative setting. Our ambient is now $X = \mathbb{G}_m^n$.

A special subvariety $T = T[\alpha]$ is determined by a finite set of (independent) multiplicative equations

$$x_1^{a_1} \cdots x_n^{a_n} = \zeta_a, \quad a = (a_1, \ldots, a_n) \in A,$$

where ζ_a is a root of unity and the lattice generated by the exponent vectors $a \in A$ is primitive.

A weakly special subvariety is determined by some additional finite set of (independent) multiplicative equations

$$x^b = x_1^{b_1} \cdots x_n^{b_n} = c_b \in \mathbb{C}^\times, \quad b = (b_1, \ldots, b_n) \in B,$$

where the lattice generated by $A \cup B$ is primitive.

A family of weakly special subvarieties of $T[\alpha]$ is then given by the set B of exponent vectors. This determines a map

$$T[\alpha] \to \mathbb{G}_m^{\#B}, \quad x \mapsto (x^b)_{b \in B}$$

with image $S[\alpha, B]$ special.

We write $T[\alpha, B]$ for the graph of this map. It determines a family of weakly special subvarieties of $T[\alpha]$ given by the fibres $T[\alpha, B]_c, c \in S[\alpha, B]$, which have dimension $\dim T[\alpha] - \#B$.

We introduce an ordering on families of weakly special subvarieties. We write $T[\alpha_1, B_1] < T[\alpha_2, B_2]$ if $T[\alpha_1]$ is a proper special subvariety of $T[\alpha_2]$, or if $T[\alpha_1] = T[\alpha_2]$ and the set $A \cup B_1$ of exponent vectors associated to equations for fibres of $T[\alpha_1, B_1]$ generates a lattice that is a proper superset of the lattice generated by $A \cup B_2$.

24.7 Theorem *Assume ZP holds for $X = \mathbb{G}_m^n$ for all n. Then $\mathrm{Opt}^0(V)$ is finite uniformly in families $V \subset X$.*

Proof About the same but interestingly a bit different! We proceed by induction on $\dim P$ and then on the ordering of families of weakly special subvarieties.

We let $V \subset X \times P$ be a family of fibres of dimension v, with each fibre V_p contained in some weakly special subvariety $T[\alpha, B]_s$ from a family $T[\alpha, B]$. We take h (to be chosen a bit later) and consider $V^{\langle h \rangle} \subset T[\alpha, B]^h$. We have $V^{\langle h \rangle} \subset T[\alpha, B]^{\langle h \rangle}$, and $T^{\langle h \rangle}$ is special. It is defined by equations of the form

$$z_i^a = \zeta_a, \quad \text{for each } a \in A \text{ and each } i,$$

and

$$z_i^b = z_j^b, \quad \text{for each } b \in B \text{ and all } i, j \; i \neq j$$

so that

$$\mathrm{codim}\big(T[\alpha, B]^{\langle h \rangle}, T[\alpha, B]^h\big) = \#B \cdot (h - 1).$$

As before, if h is sufficiently large, any h optimal points on a fibre V_p leads to a point $z = (z_1, \ldots, z_h) \in V^{\langle h \rangle}$ that is atypical for $V^{\langle h \rangle}$ in $T[\alpha, B]^{\langle h \rangle}$, and so z satisfies one of finitely many multiplicative relations of the form

$$\prod z_{ij}^{c_{ij}} = \zeta_c, \quad c \in C.$$

Fix $c \in C$. We look at the exponent vector c_i on a particular $z_i \in X$. If $c_i \notin \mathbb{Z}\langle A, B \rangle$, then if we have sufficiently many points $z_i^{(\ell)}$ and points z_j, $j \neq i$, such that all the points

$$(z_1, \ldots, z_{i-1}, z_i^{(\ell)}, z_{i+1}, \ldots, z_n), \quad \text{all } \ell,$$

satisfy $z^c = \zeta_c$, then we get many points $z_i^{(\ell)}$ on a fibre of family of weakly special subvarieties with a larger set of fixed exponents $B \cup \{c\}$.

If, on the other hand, every such c_i is in $\mathbb{Z}\langle A, B \rangle$, then we may (using the relations on $T^{\langle h \rangle}$) write $z^c = \zeta_c$ equivalently as a special condition on one set of coordinates z_1, say, and it must be a relation that does not hold on $[T]$.

Thus, by the Hypergraph Ramsey Theorem, if many points satisfy one of finitely many such conditions, we end up with many points on a fibre of one of finitely many smaller (in our ordering) families of weakly special subvarieties.

The proof is thus complete by induction.   □

24.8 Proof of Theorem 24.2 in the Multiplicative Case

This follows from Proposition 24.3 and Theorem 24.7.   □

References

[1] B. Allombert, Y. Bilu, and A. Pizarro-Madariaga, CM points on straight lines, *Analytic number theory in honour of Helmut Maier's 60th birthday, 1–18,* C. Pomerance and M. Rassias, eds., Springer, Cham, 2015.

[2] L. V. Ahlfors, *Complex analysis,* 3rd ed., McGraw-Hill, New York, 1979.

[3] M. Amou, On algebraic independence of certain functions related to the elliptic modular function, *Number theory and its applications (Kyoto, 1997),* 25–34, Dev. Math. **2**, Shigeru Kanemitsu and Kalman Györy, eds., Kluwer Academic, Dordrecht, 1999.

[4] Y. André, *G-functions and geometry,* Aspects of Mathematics **E13**, Vieweg, Braunschweig, 1989.

[5] Y. André, Mumford-Tate groups of mixed Hodge structures and the theorem of the fixed part, *Compos. Math.* **82** (1992), 1–24.

[6] Y. André, Finitude des couples d'invariants modulaire singuliers sur une courbe algébrique plane non modulaire, *J. Reine Angew. Math.* **505** (1998), 203–208.

[7] Y. André, Shimura varieties, subvarieties, and CM points, six lectures at the Franco-Taiwan arithmetic festival, August–September 2001.

[8] Y. André, *Une introduction aux motifs (motifs purs, motifs mixtes, périodes),* Panoramas et Synthèses **17**, Société Mathématique de France, Paris, 2004.

[9] Y. André, Letter to C. Bertolin dated 29 May 2019, reprinted in [48].

[10] Y. André, P. Corvaja, and U. Zannier (with an appendix by Z. Gao), The Betti map associated to a section of an abelian scheme, *Invent. Math.* **222** (2020), 161–202.

[11] F. Andreatta, E. Goren, B. Howard, and K. Madapusi Pera, Faltings height of abelian varieties with complex multiplication, *Ann. Math.* **187** (2018), 391–531.

[12] J. Armitage, Pfaffian control of some polynomials involving the *j*-function and Weierstrass elliptic functions, arXiv:2011.09382.

[13] V. Aslanyan, Ax–Schanuel type theorems and geometry of strongly minimal sets in differentially closed fields, arXiv:1606.01778.

[14] V. Aslanyan, Weak modular Zilber–Pink with derivatives, *Math. Ann.* (2021). https://doi.org/10.1007/s00208-021-02213-7

[15] V. Aslanyan, Some remarks on atypical intersections, *Proc. Am. Math. Soc.* **139** (2021), 4649–4660.

[16] V. Aslanyan, Ax–Schanuel and strong minimality for the *j*-function, *Ann. Pure Appl. Logic* **172** (2021), paper 102871.

214

[17] V. Aslanyan, S. Eterovic, and J. Kirby, Differential existential closedness for the *j*-function, *Proc. Am. Math. Soc.* **149** (2021), 1417–1429.

[18] V. Aslanyan, S. Eterovic, and J. Kirby, A closure operator respecting the modular *j*-function, *Isr. J. Math.*, to appear.

[19] V. Aslanyan, J. Kirby, and V. Mantova, A geometric approach to some systems of exponential equations, arXiv:2105.12679.

[20] J. Ax, On Schanuel's conjectures, *Ann. Math.* **93** (1971), 252–268.

[21] J. Ax, Some topics in differential algebraic geometry I: Analytic subgroups of algebraic groups, *Am. J. Math.* **94** (1972), 1195–1204.

[22] J. Ayoub, Une version relative de la conjecture des périodes de Kontsevich-Zagier, *Ann. Math.* **181** (2015), 905–992.

[23] J. Ayoub, Periods and the conjectures of Grothendieck and Kontsevich-Zagier, *Newsl. Eur. Math. Soc.* **91** (March 2014), 12–18.

[24] A. Baker, *Transcendental number theory,* Cambridge University Press, Cambridge, 1975/1979.

[25] M. Baker and L. De Marco, Special curves and postcritically finite polynomials, *Forum Math. Pi* **1** (2013), e3, 35pp.

[26] B. Bakker, Y. Brunebarbe, B. Klingler, and J. Tsimerman, Definability of mixed period maps, arXiv:2006.12403, *J. Eur. Math. Soc.,* to appear.

[27] B. Bakker, Y. Brunebarbe, and J. Tsimerman, O-minimal GAGA and a conjecture of Griffiths, arXiv:1811.12230v2.

[28] B. Bakker, Y. Brunebarbe, and J. Tsimerman, Quasiprojectivity of images of mixed period maps, arXiv:2006.13709.

[29] B. Bakker, B. Klingler, and J. Tsimerman, Tame topology of arithmetic quotients and algebraicity of Hodge loci, *J. Am. Math. Soc.* **33** (2020), 917–939.

[30] B. Bakker and J. Tsimerman, The Ax–Schanuel conjecture for variations of Hodge structure, *Invent. Math.* **217** (2019), 77–94.

[31] B. Bakker and J. Tsimerman, Lectures on the Ax–Schanuel conjecture, *Arithmetic geometry of logarithmic pairs and hyperbolicity of moduli spaces*, 1–68, M.-H. Nicole, ed., CRM Short Courses, Springer, Cham, 2020.

[32] G. Baldi, On a conjecture of Buium and Poonen, *Ann. Inst. Fourier* **70** (2020), 457–477.

[33] G. Baldi and E. Ullmo, Special subvarieties of non-arithmetic ball quotients and Hodge theory, arXiv:2005.03524.

[34] F. Barroero, CM relations in fibred powers of elliptic families, *J. Inst. Math. Jussieu* **18** (2019), corrigendum, *J. Inst. Math. Jussieu* (2018), online.

[35] F. Barroero and L. Capuano, Linear relations in families of powers of elliptic curves, *Algebra Number Theory* **10** (2016), 195–214.

[36] F. Barroero and L. Capuano, Unlikely intersections in products of families of elliptic curves and the multiplicative group, *Q. J. Math.* **68** (2017), 1117–1138.

[37] F. Barroero and L. Capuano, Unlikely intersections in families of abelian varieties and the polynomial Pell equation, *Proc. London Math. Soc.* **120** (2020), 192–219.

[38] F. Barroero, L. Capuano, L. Mérai, A. Ostafe, and M. Sha, Multiplicative and linear dependence in finite fields and on elliptic curves modulo primes, arXiv:2008.00389.

[39] F. Barroero and G. A. Dill, On the Zilber–Pink conjecture for complex abelian varieties, *Ann. Sci. Ec. Norm. Super.*, to appear.

[40] F. Barroero and G. A. Dill, Distinguished categories and the Zilber–Pink conjecture, arXiv:2103.07422.

[41] F. Barroero, L. Kühne, and H. Schmidt, Unlikely intersections of curves with algebraic subgroups in semiabelian varieties, arXiv:2108.12405.

[42] F. Barroero and M. Sha, Torsion points with multiplicatively dependent coordinates on elliptic curves, *Bull. London Math. Soc.* **52** (2020), 807–815.

[43] F. Barroero and M. Widmer, Counting lattice points and o-minimal structures, *Int. Math. Res. Not.* **2014**, 4932–4957.

[44] M. Bays and P. Habegger, A note on divisible points on curves, *Trans. Am. Math. Soc.* **367** (2015), 1313–1328.

[45] M. Bays, J. Kirby, and A. J. Wilkie, A Schanuel property for exponentially transcendental powers, *Bull. London Math. Soc.* **42** (2010), 917–922.

[46] A. Berarducci and T. Servi, An effective version of Wilkie's theorem of the complement and some effective o-minimality results, *Ann. Pure Appl. Logic* **125** (2004), 43–74.

[47] C. Bertolin, Périodes de 1-motifs et transcendance, *J. Number Theory* **97** (2002), 204–221.

[48] C. Bertolin, Third kind elliptic integrals and 1-motives, *J. Pure Appl. Algebra* **224** (2020), 106396, 28pp.

[49] D. Bertrand, Theta functions and transcendence, *Ramanujan J.* **1** (1997), 339–350.

[50] D. Bertrand, Schanuel's conjecture for non-isoconstant elliptic curves over function fields, *Model theory and its applications to algebra and analysis, volume 1*, 41–62, Z. Chatzidakis, D. Macpherson, A. Pillay, and A. J. Wilkie, eds., LMS Lecture Note Series **349**, Cambridge University Press, Cambridge, 2008.

[51] D. Bertrand, Théorie de Galois différentielles et transcendence, *Ann. Inst. Fourier* **59** (2009), 2773–2803.

[52] D. Bertrand, Unlikely intersections in Poincaré biextensions over elliptic schemes, *Notre Dame J. Formal Logic* **54** (2013), 365–375.

[53] D. Bertrand, Special points and Poincaré bi-extensions, with an appendix by S. J. Edixhoven, arXiv:1104.5178.

[54] D. Bertrand and S. J. Edixhoven, Pink's conjecture on unlikely intersections and families of semi-abelian varieties, *J. Éc. Polytech. Math.* **7** (2020), 711–742.

[55] D. Bertrand, D. Masser, A. Pillay, and U. Zannier, Relative Manin–Mumford for semi-abelian surfaces, *Proc. Edinburgh Math. Soc.* **59** (2016), 837–875.

[56] D. Bertrand and A. Pillay, A Lindemann–Weierstrass theorem for semi-abelian varieties over function fields, *J. Am. Math. Soc.* **23** (2010), 491–533.

[57] D. Bertrand and H. Schmidt, Unlikely intersections in semi-abelian surfaces, *Algebra Number Theory* **13** (2019), 1455–1473.

[58] D. Bertrand and W. Zudilin, On the transcendence degree of the differential field generated by Siegel modular functions, *J. Reine Angew. Math.* **554** (2003), 47–68.

[59] E. Besson, Points rationnels de la fonction gamma d'Euler, *Arch. Math. (Basel)* **103** (2014), 61–73.

[60] N. Bhardwaj and L. van den Dries, On the Pila-Wilkie Theorem, arXiv:2010.14046.

[61] E. Bierstone and P. D. Milman, Semianalytic and subanalytic sets, *Publ. Math. Inst. Hautes Etudes Sci.* **67** (1988), 5–42.

[62] Y. Bilu, Limit distribution of small points on algebraic tori, *Duke Math. J.* **89** (1997), 465–476.

[63] Y. Bilu, P. Habegger, and L. Kühne, No singular modulus is a unit, *Int. Math. Res. Not.* **2020**, 10005–10041.

[64] Y. Bilu and L. Kühne, Linear equations in singular moduli, *Int. Math. Res. Not.* **2020**, 7617–7643.

[65] Y. Bilu, F. Luca, and D. Masser, Collinear CM points, *Algebra Number Theory* **11** (2017), 1047–1087.

[66] Y. Bilu, F. Luca, and A. Pizarro-Madariaga, Rational products of singular moduli, *J. Number Theory* **158** (2016), 397–410.

[67] Y. Bilu, D. Masser, and U. Zannier, An effective "theorem of André" for CM points on plane curves, *Math. Proc. Cambridge Philos. Soc.* **154** (2013), 145–152.

[68] G. Binyamini, Zero-counting and invariant sets for differential equations, *Int. Math. Res. Not.* **2019**, 4119–4158.

[69] G. Binyamini, Density of algebraic points on Noetherian varieties, *Geom. Funct. Anal.* **29** (2019), 72–118.

[70] G. Binyamini, Some effective estimates for André–Oort in $Y(1)^n$, with an appendix by E. Kowalski, *J. Reine Angew. Math.* **767** (2020), 17–35.

[71] G. Binyamini, Point counting for foliations over number fields, arXiv:2009.00892.

[72] G. Binyamini, R. Cluckers, and D. Novikov, Point-counting and Wilkie's conjecture for non-archimedean Pfaffian and Noetherian functions, *Duke Math. J.*, to appear.

[73] G. Binyamini and C. Daw, Effective computations for weakly optimal subvarieties, arXiv:2105.12760.

[74] G. Binyamini and D. Masser, Effective André–Oort for non-compact curves in Hilbert modular varieties, *C. R. Acad. Sci. Paris, Ser. I* **359** (2021), 313–321.

[75] G. Binyamini and D. Novikov, The Pila-Wilkie theorem for subanalytic families: A complex analytic approach, *Compos. Math.* **153** (2017), 2171–2194.

[76] G. Binyamini and D. Novikov, Wilkie's conjecture for restricted elementary functions, *Ann. Math.* **186** (2017), 237–275.

[77] G. Binyamini and D. Novikov, Complex cellular structures, with an appendix by Y. Yomdin, *Ann. Math.* **190** (2019), 145–248.

[78] G. Binyamini and D. Novikov, The Yomdin-Gromov algebraic lemma revisited, *Arnold Math. J.* **7** (2021), 419–430.

[79] G. Binyamini, H. Schmidt, and A. Yafaev, Lower bounds for Galois orbits of special points on Shimura varieties: A point-counting approach, arXiv:2104.05842, *Math. Ann.*, to appear.

[80] G. Binyamini and N. Vorobjov, Effective cylindrical cell decompositions for restricted sub-Pfaffian sets, *Int. Math. Res. Not.* **2022**, 3493–3510.

[81] C. Birkenhake and H. Lange, *Complex abelian varieties,* Grund. der math. Wiss. **302**, 2nd ed., Springer-Verlag, Berlin, 2004.

[82] D. Blázquez-Sanz, G. Casale, J. Freitag, and J. Nagloo, Some functional transcendence results around the Schwarzian differential equation, arXiv:1912.09963.

[83] D. Blázquez-Sanz, G. Casale, J. Freitag, and J. Nagloo, A differential approach to the Ax–Schanuel, I, arXiv:2102.03384.

[84] F. Bogomolov, Points of finite order on an abelian variety, *Math. USSR-Izv.* **17** (1981), 55–72.

[85] E. Bombieri and W. Gubler, *Heights in diophantine geometry,* Cambridge University Press, Cambridge, 2006.

[86] E. Bombieri, P. Habegger, D. Masser, and U. Zannier, A note on Maurin's theorem, *Rend. Lincei-Mat. Appl.* **21** (2010), 251–260.

[87] E. Bombieri, D. Masser, and U. Zannier, Intersecting a curve with algebraic subgroups of multiplicative groups, *Int. Math. Res. Not.* **1999**, 1119–1140.

[88] E. Bombieri, D. Masser, and U. Zannier, Finiteness results for multiplicatively dependent points on complex curves, *Michigan Math. J.* **51** (2003), 451–466.

[89] E. Bombieri, D. Masser, and U. Zannier, Intersecting curves and algebraic subgroups: Conjectures and more results, *Trans. Am. Math. Soc.* **358** (2006), 2247–2257.

[90] E. Bombieri, D. Masser, and U. Zannier, Anomalous subvarieties – structure theorems and applications, *Int. Math. Res. Not.* **2007**, Art. ID rnm057, 33pp.

[91] E. Bombieri, D. Masser, and U. Zannier, Intersecting a plane with algebraic subgroups of multiplicative groups, *Ann. Sc. Norm. Super. Pisa Cl. Sci. (5)* **7** (2008), 51–80.

[92] E. Bombieri, D. Masser, and U. Zannier, On unlikely intersections of complex varieties with tori, *Acta Arith.* **133** (2008), 309–323.

[93] E. Bombieri and J. Pila, The number of integral points on arcs and ovals, *Duke Math. J.* **59** (1989), 337–357.

[94] E. Bombieri and U. Zannier, Heights of algebraic points on subvarieties of abelian varieties, *Ann. Sc. Norm. Super. Pisa Cl. Sci. (4)* **23** (1996), 779–792.

[95] M. Boshernitzan, Hardy fields and existence of transexponential functions, *Aequ. Math.* **30** (1986), 258–280.

[96] G. Boxall, A special case of quasiminimality, *Q. J. Math.* **71** (2020), 1065–1068.

[97] G. Boxall, T. Chalebgwa, and G. Jones, On algebraic values of Weierstrass σ-functions, arXiv:2011.11980.

[98] G. Boxall and G. O. Jones, Algebraic values of certain analytic functions, *Int. Math. Res. Not.* **2015**, 1141–1158.

[99] G. Boxall and G. O. Jones, Rational values of entire functions of finite order, *Int. Math. Res. Not.* **2015**, 12,251–12,264.

[100] G. Boxall, G. O. Jones, and H. Schmidt, Rational values of transcendental functions and arithmetic dynamics, *J. Eur. Math. Soc.* (2018). http://arxiv.org/abs/1808.07676

[101] F. Breuer, Heights of CM points on complex affine curves, *Ramanujan J.* **5** (2001), 311–317.

[102] F. Breuer, Special subvarieties of Drinfeld modular varieties, *J. Reine Angew. Math.* **668** (2012), 35–57.

[103] P. Brosnan, Volumes of definable sets in o-minimal structures and affine GAGA theorems, arXiv:2102.01227.

[104] W. D. Brownawell and K. Kubota, The algebraic independence of Weierstrass functions and some related numbers, *Acta Arith.* **33** (1977), 111–149.

[105] W. D. Brownawell and D. W. Masser, Zero estimates with moving targets, *J. London Math. Soc.* **95** (2017), 441–454.

[106] W. D. Brownawell and D. W. Masser, Unlikely intersections for curves in additive groups over positive characteristic, *Proc. Am. Math. Soc.* **145** (2017), 4617–4627.

[107] T. D. Browning and D. R. Heath-Brown, Plane curves in boxes and equal sums of two powers, *Math. Z.* **251** (2005), 233–247.

[108] T. D. Browning, D. R. Heath-Brown, and P. Salberger, Counting rational points on algebraic varieties, *Duke Math. J.* **132** (2006), 545–578.

[109] A. Buium, Geometry of differential polynomial functions, III: Moduli spaces, *Am. J. Math.* **117** (1995), 1–73.

[110] A. Buium and B. Poonen, Independence of points on elliptic curves arising from special points on modular and Shimura curves, I: Global results, *Duke Math. J.* **147** (2009), 181–191.

[111] D. Burguet, A proof of Yomdin-Gromov's algebraic lemma, *Isr. J. Math.* **168** (2008), 291–316.

[112] D. Burguet, G. Liao, and J. Yang, Asymptotic h-expansiveness rate of C^∞ maps, *Proc. London Math. Soc.* **111** (2015), 381–419.

[113] L. Butler, Some cases of Wilkie's conjecture, *Bull. London Math. Soc.* **44** (2012), 642–660.

[114] L. Butler, A diophantine approach to the three and four exponentials conjectures, *Ramanujan J.* **42** (2017), 199–221.

[115] A. Cadoret and A. Kret, Galois-generic points on Shimura varieties, *Algebra Number Theory* **10** (2016), 1893–1934.

[116] F. Campagna, On singular moduli that are S-units, arXiv:1904.08958.

[117] V. Cantoral-Farfan, K. H. Nguyen, and F. Vermeulen, A Pila-Wilkie theorem for Hensel minimal curves, arXiv:2107.03643.

[118] L. Capuano, D. Masser, J. Pila, and U. Zannier, Rational points on Grassmannians and unlikely intersections in tori, *Bull. London Math. Soc.* **48** (2016), 141–154.

[119] M. Carrizosa, Problème de Lehmer et variétés abéliennes CM, *C. R. Math. Acad. Sci. Paris* **346** (2008), 1219–1224.

[120] M. Carrizosa, Petits points et multiplication complexe, *Int. Math. Res. Not.* **2009**, 3016–3097.

[121] G. Casale, J. Freitag, and J. Nagloo, Ax–Lindemann-Weierstrass with derivatives and the genus 0 Fuchsian groups, *Ann. Math.* **192** (2020), 721–765.

[122] O. Cassani, The defect condition, MODNET eprint 1936, March 2021.

[123] W. Castryck, R. Cluckers, P. Dittmann, and K. H. Nguyen, The dimension growth conjecture, polynomial in the degree and without logarithmic factors, *Algebra Number Theory* **14** (2020), 2261–2294.

[124] C.-L. Chai and F. Oort, Abelian varieties isogenous to a Jacobian, *Ann. Math.* **176** (2012), 589–635.

[125] T. P. Chalebgwa, Algebraic values of certain analytic functions defined by a canonical product, *Bull. Aust. Math. Soc.* **101** (2020), 415–425.

[126] T. P. Chalebgwa, Nevanlinna theory and algebraic values of certain meromorphic functions, *Int. J. Number Theory* **16** (2020), 1607–1636.

[127] T. P. Chalebgwa, Algebraic values of entire functions with extremal growth orders: An extension of a theorem of Boxall and Jones, *Can. Math. Bull.* **63** (2020), 536–546.

[128] A. Chambert-Loir, Relations de dépendance et intersections exceptionnelles, Séminaire Bourbaki (2010–2011), exposé 1032.

[129] A. Chambert-Loir and F. Loeser, A non-archimedean Ax–Lindemann theorem, *Algebra Number Theory* **11** (2017), 1967–1999.

[130] J.-Y. Charbonnel, Sur certains sous-ensembles de l'espace euclidien, *Ann. Inst. Fourier* **41** (1991), 679–717.

[131] Z. Chatzidakis, D. Ghioca, D. Masser, and G. Maurin, Unlikely, likely and impossible intersections without algebraic groups, *Atti Acad. Naz. Lincei Rend. Lincei-Mat. Appl.* **24** (2013), 485–501.

[132] S. Checcoli, F. Veneziano, and E. Viada, On torsion anomalous intersections, *Atti Accad. Naz. Lincei Rend. Lincei Mat. App.* **25** (2014), 1–36.

[133] S. Checcoli, F. Veneziano, and E. Viada, The explicit Mordell conjecture for families of curves (with an appendix by M. Stoll), *Forum Math. Sigma* **7** (2019), e31, 62pp.

[134] S. Checcoli and E. Viada, On the torsion anomalous conjecture in CM abelian varieties, *Pac. J. Math.* **271** (2014), 321–345.

[135] I. Chen and G. Glebov, On Chudnovsky–Ramanujan type formulae, *Ramanujan J.* **46** (2018), 677–712.

[136] J. Chen, On the geometric André–Oort conjecture for variations of Hodge structures, arXiv:2010.09643.

[137] K. C. T. Chiu, Ax–Schanuel for variations of mixed Hodge structures, arXiv:2101.11593.

[138] K. C. T. Chiu, Ax–Schanuel with derivatives for mixed period mappings, arXiv:2110.03489.

[139] C. Ciliberto and G. van der Geer, Subvarieties of the moduli space of curves parameterizing Jacobians with non-trivial endomorphisms, *Am. J. Math.* **114** (1992), 551–570.

[140] L. Clozel, H. Oh, and E. Ullmo, Hecke operators and equidistribution of Hecke points, *Invent. Math.* **144** (2001), 327–351.

[141] L. Clozel and E. Ullmo, Equidistribution de sous-variétés spéciales, *Ann. Math.* **161** (2005), 1571–1588.

[142] R. Cluckers, G. Comte, and F. Loeser, Non-Archimedean Yomdin-Gromov parameterizations and points of bounded height, *Forum Math. Pi* **3** (2015), e5, 60pp.

[143] R. Cluckers, A. Forey, and F. Loeser, Uniform Yomdin-Gromov parameterizations and points of bounded height in valued fields, *Algebra Number Theory* **14** (2020), 1423–1456.

[144] R. Cluckers, I. Halupczok, and S. Rideau-Kikuchi, Hensel minimality I, arXiv:1909.13792.

[145] R. Cluckers, I. Halupczok, S. Rideau-Kikuchi, and F. Vermeulen, Hensel minimality II: Mixed characteristic and a diophantine application, arXiv:2104.09475.

[146] R. Cluckers, J. Pila, and A. J. Wilkie, Uniform parameterization of subanalytic sets and diophantine applications, *Ann. Sci. Ec. Norm. Super.* **53** (2020), 1–42.

[147] P. B. Cohen (now Tretkoff), On the coefficients of the transformation polynomials for the elliptic modular function, *Math. Proc. Cambridge Philos Soc.* **95** (1984), 389–402.

[148] P. B. Cohen (now Tretkoff), Humbert surfaces and transcendence properties of automorphic functions, *Rocky Mt. J. Math.* **26** (1996), 987–1001.

[149] P. B. Cohen (now Tretkoff), Hyperbolic distribution problems on Siegel 3-folds and Hilbert modular varieties, *Duke Math. J.* **129** (2005), 87–127.

[150] P. B. Cohen (now Tretkoff) and U. Zannier, Fewnomials and intersections of lines with real analytic subgroups in $\mathbb{G}_m^n$, *Bull. London Math. Soc.* **34** (2002), 21–32.

[151] P. J. Cohen, Decision procedures for real and p-adic fields, *Commun Pure Appl. Math.* **22** (1969), 131–151.

[152] P. Colmez, Périodes des variétés abéliennes à multiplications complexe, *Ann. Math.* **138** (1993), 625–683.

[153] J. Commelin, P. Habegger, and A. Huber, Exponential periods and o-minimality, I, arXiv:2007.08280.

[154] J. Commelin and A. Huber, Exponential periods and o-minimality, II, arXiv:2007.08290.

[155] G. Comte and C. Miller, Points of bounded height on oscillatory sets, *Q. J. Math.* **68** (2017), 1261–1287.

[156] G. Comte and Y. Yomdin, Zeroes and rational points of analytic functions, *Ann. Inst. Fourier* **68** (2018), 2445–2476.

[157] J. H. Conway and A. J. Jones, Trigonometric Diophantine equations (On vanishing sums of roots of unity), *Acta Arith.* **30** (1976), 229–240.

[158] P. Corvaja, D. Masser, and U. Zannier, Sharpening "Manin–Mumford" for certain algebraic groups of dimension 2, *Enseign. Math.* **59** (2013), 225–269.

[159] P. Corvaja, D. Masser, and U. Zannier, Torsion hypersurfaces on abelian schemes and Betti coordinates, *Math. Ann.* **371** (2018), 1013–1045.

[160] P. D'Aquino, A. Fornasiero, and G. Terzo, Generic solutions of equations with iterated exponentials, *Trans. Am. Math. Soc.* **370** (2018), 1393–1407.

[161] P. D'Aquino, A. Macintyre, and G. Terzo, Schanuel's nullstellensatz for Zilber fields, *Fundam Math.* **207** (2010), 123–143.

[162] C. Daw, Degrees of strongly special subvarieties and the André–Oort conjecture, *J. Reine Angew. Math.* **721** (2016), 81–108.

[163] C. Daw, The André–Oort conjecture via o-minimality, *O-minimality and diophantine geometry,* 129–158, G. O. Jones and A. J. Wilkie, eds., LMS Lecture Notes Series **421**, Cambridge University Press, Cambridge, 2015.

[164] C. Daw and A. Harris, Categoricity of modular and Shimura curves, *J. Inst. Math. Jussieu* **16** (2017), 1075–1101.

[165] C. Daw, A. Javanpeykar, and L. Kühne, Effective estimates for the degrees of maximal special subvarieties, *Sel. Math. New Ser.* **26** (2020), No. 1, Art. 2, 31pp.

[166] C. Daw and M. Orr, Heights of pre-special points of Shimura varieties, *Math. Ann.* **365** (2016), 1305–1357.

[167] C. Daw and M. Orr, Unlikely intersections with $E \times$ CM curves in $\mathcal{A}_2$, *Ann. Sc. Norm. Super. Pisa Cl. Sci.*, **22** (2021), no. 4, 1705–1745

[168] C. Daw and M. Orr, Quantitative reduction theory and unlikely intersections, *Int. Math. Res. Not.*, **2021**, rnab173, https://doi.org/10.1093/imrn/rnab173

[169] C. Daw and J. Ren, Applications of the hyperbolic Ax–Schanuel conjecture, *Compos. Math.* **154** (2018), 1843–1888.

[170] C. Daw and A. Yafaev, An unconditional proof of the André–Oort conjecture for Hilbert modular surfaces, *Manuscr. Math.* **135** (2011), 263–271.

[171] P. Deligne, Travaux de Shimura, Séminaire Bourbaki, 23ème année (1970–71), exposé 389, *Lecture Notes in Mathematics* **244**, 123–165, Springer, Berlin, 1971.

[172] P. Deligne, Théorie de Hodge: II, *Publ. Math. Inst. Hautes Etudes Sci.* **40** (1972), 5–47.

[173] P. Deligne, La conjecture de Weil pour les surface $K3$, *Invent. Math.* **15** (1972), 206–226.

[174] P. Deligne, Variétés de Shimura: Interprétation modulaire, et techniques de construction de modèles canoniques, *Automorphic forms, representations, and L-functions (Corvalis, 1977), volume 2*, 247–289, Proc. Symp. Pure Math. **XXXIII**, A. Borel and W. Casselman, eds., AMS, Providence, RI, 1979.

[175] L. DeMarco, H. Krieger, and H. Ye, Uniform Manin–Mumford for a family of genus 2 curves, *Ann. Math.* **191** (2020), 949–1001.

[176] L. DeMarco, X. Wang, and H. Ye, Torsion points and the Lattés family, *Am. J. Math.* **138** (2016), 697–732.

[177] J. Denef and L. van den Dries, p-adic and real subanalytic sets, *Ann. Math.* **128** (1988), 79–138.

[178] F. Diamond and J. Shurman, *A first course in modular forms,* Graduate texts in mathematics **228**, Springer-Verlag, New York, 2005.

[179] G. A. Dill, Unlikely intersections between isogeny orbits and curves, *J. Eur. Math. Soc.* **23** (2021), 2405–2438.

[180] G. A. Dill, Unlikely intersections with isogeny orbits in a product of elliptic schemes, *Math. Ann.* **377** (2020), 1509–1545.

[181] G. A. Dill, Torsion points on isogenous abelian varieties, arXiv:2011.05815.

[182] V. Dimitrov, Z. Gao, and P. Habegger, Uniformity in Mordell–Lang for curves, *Ann. Math.* **194** (2021), 237–298.

[183] T.-C. Dinh and D.-V. Vu, Algebraic flows on commutative complex lie groups, arXiv:1811.09823.

[184] N. Dogra, Unlikely intersections and the Chabauty-Kim method over number fields, arXiv:1903.05032.

[185] I. Dolgachev, *Endomorphisms of abelian varieties,* Lecture Notes, Milan, 2014, available from the author's home page.

[186] L. van den Dries, Remarks on Tarski's problem concerning $(\mathbb{R}, +, \cdot, \exp)$, *Logic colloquium '82,* 97–121, G. Lolli, G. Longo, and A. Marcja, eds., Studies in Logic and the Foundations of Mathematics **112**, North Holland, Amsterdam, 1984.

[187] L. van den Dries, A generalization of the Tarski-Seidenberg theorem, and some non-definability results, *Bull. Am. Math. Soc.* **15** (1986), 189–193.

[188] L. van den Dries, On the elementary theory of restricted elementary functions, *J. Symb. Log.* **53** (1988), 796–808.

[189] L. van den Dries, *Tame topology and o-minimal structures,* LMS Lecture Note Series **248**, Cambridge University Press, Cambridge, 1998.

[190] L. van den Dries and A. Günaydin, The fields of real and complex numbers with a small multiplicative group, *Proc. London Math. Soc.* **93** (2003), 43–81.

[191] L. van den Dries, A. Macintyre, and D. Marker, The elementary theory of restricted analytic fields with exponentiation, *Ann. Math.* **140** (1994), 183–205.

[192] L. van den Dries and C. Miller, On the real exponential field with restricted analytic functions, *Isr. J. Math.* **85** (1994), 19–56.

[193] L. van den Dries and C. Miller, Geometric categories and o-minimal structures, *Duke Math. J.* **84** (1996), 497–540.

[194] W. Duke, Hyperbolic distribution problems and half-integral weight Maass forms, *Invent. Math.* **92** (1988), 73–90.

[195] R. Dvornicich and U. Zannier, On sums of roots of unity, *Monatsh. Math.* **129** (2000), 97–108.

[196] S. J. Edixhoven, Special points on the product of two modular curves, *Compos. Math.* **114** (1998), 315–328.

[197] S. J. Edixhoven, On the André–Oort conjecture for Hilbert modular surfaces, *Moduli of abelian varieties (Texel Island, 1999)*, 133–155, Progress in Mathematics **195**, C. Faber, G. van der Geer and F. Oort, eds., Birkhauser, Basel, 2001.

[198] S. J. Edixhoven, Special points on products of modular curves, *Duke Math. J.* **126** (2005), 325–348.

[199] S. J. Edixhoven, B. J. J. Moonen, and F. Oort, Open problems in algebraic geometry, *Bull. Sci. Math.* **125** (2001), 1–22.

[200] S. J. Edixhoven and R. Richard, A mod p variant of the André–Oort conjecture, *Rend. Circ. Mat. Palermo* **69** (2020), 151–157.

[201] S. J. Edixhoven and A. Yafaev, Subvarieties of Shimura varieties, *Ann. Math.* **157** (2003), 621–645.

[202] M. Einsiedler, G. Margulis, and A. Venkatesh, Effective equidistribution for closed orbits of semisimple groups on homogeneous spaces, *Invent. Math.* **177** (2009), 137–212.

[203] P. Eleftheriou, Counting algebraic points in expansions of o-minimal structures by a dense set, *Q. J. Math.* **72** (2021), 817–833.

[204] N. Elkies, Rational points near curves and small non-zero $|x^3 - y^2|$ via lattice reduction, *Algorithmic number theory (Leiden, 2000)*, 33–63, Lecture Notes in Comp. Sci. **1838**, Wieb Bosma, ed. Springer, Berlin, 2000.

[205] J. S. Ellenberg, C. Elsholtz, C. Hall, and E. Kowalski, Non-simple abelian varieties in a family: Geometric and analytic approaches, *J. London Math. Soc.* **80** (2009), 135–154.

[206] J. S. Ellenberg, B. Lawrence, and A. Venkatesh, Sparsity of integral points on moduli spaces of varieties, arXiv:2109.01043.

[207] J. S. Ellenberg and A. Venkatesh, On uniform bounds for rational points on non-rational curves, *Int. Math. Res. Not.* **2005**, 2163–2181.

[208] S. Eterovic, A Schanuel property for j, *Bull. London Math. Soc.* **50** (2018), 293–315.

[209] S. Eterovic, Categoricity of Shimura varieties, arXiv:1803.06700.

[210] S. Eterovic and S. Herrero, Solutions of equations involving the modular j function, *Trans. Am. Math. Soc.* **374** (2021), 3971–3998.

[211] S. Eterovic and R. Zhao, Algebraic varieties and automorphic functions, arXiv:2107.10392.

[212] G. Faltings, Endlichkeitssätze für abelsche Varietäten über Zahlkörpern, *Invent. Math.* **73** (1983), 349–366.

[213] G. Faltings, Diophantine approximation on abelian varieties, *Ann. Math.* **133** (1991), 549–576.

[214] G. Faltings, The general case of S. Lang's conjecture, *Barsotti symposium in algebraic geometry (Abano Terme, 1991)*, 175–182, Perspect. Math. **15**, V. Cristante and W. Messing, eds., Academic Press, San Diego, 1994.

[215] T. J. Fonseca, Higher Ramanujan equations and periods of abelian varieties, *Mem. Am. Math. Soc.*, to appear.

[216] G. Fowler, Triples of singular moduli with rational product, *Int. J. Number Theory* **16** (2020), 2149–2166.

[217] G. Fowler, Multiplicative independence of modular functions, arXiv:2005.13328.

[218] G. Fowler, Equations in three singular moduli: The equal exponent case, arXiv:2105.12696.

[219] J. Fox, M. Kwan, and H. Spink, Geometric and o-minimal Littlewood-Offord problems, arXiv:2106.04894.

[220] J. Freitag, Not Pfaffian, arXiv:2109.09230.

[221] J. Freitag and T. Scanlon, Strong minimality and the *j*-function, *J. Eur. Math. Soc.* **20** (2018), 119–136.

[222] A. Gabrielov, Projections of semi-analytic sets, *Funct. Anal. Appl.* **2** (1968), 282–291.

[223] A. Gabrielov and N. Vorobjov, Complexity of computations with Pfaffian and Noetherian functions, *Normal forms, bifurcations and finiteness problems in differential equations,* 211–250, NATO Sci. Ser. II: Math. Phys. Chem. **137**, Y. Ilyashenko, C. Rousseau and G. Sabidussi, eds., Kluwer Academic, Dordrecht, 2004.

[224] A. Galateau, Une minoration du minimum essentiel sur les variétés abéliennes, *Comment. Math. Helv.* **85** (2010), 775–812.

[225] Z. Gao, About the mixed André–Oort conjecture: Reduction to a lower bound for the pure case, *C. R. Acad. Sci. Paris* **354** (2016), 659–663.

[226] Z. Gao, Towards the André–Oort conjecture for mixed Shimura varieties: The Ax–Lindemann theorem and lower bounds for Galois orbits of special points, *J. Reine Angew. Math.* **732** (2017), 85–146.

[227] Z. Gao, A special point problem of André-Pink-Zannier in the universal family of abelian varieties, *Ann. Sci. Norm. Sup. Pisa Cl. Sci.* **17** (2017), 231–266.

[228] Z. Gao, Enlarged mixed Shimura varieties, bi-algebraic systems, and some Ax-type transcendental results, *Forum Math. Sigma* **7** (2019), e16, 65pp.

[229] Z. Gao, Mixed Ax–Schanuel for the universal abelian varieties and some applications, *Compos. Math.* **156** (2020), 2263–2297.

[230] Z. Gao, Generic rank of Betti map and unlikely intersections, *Compos. Math.* **156** (2020), 2469–2509.

[231] Z. Gao, T. Ge, and L. Kühne, The uniform Mordell–Lang conjecture, arXiv:2105.15085.

[232] Z. Gao and P. Habegger, Heights in families of abelian varieties and the geometric Bogomolov conjecture, *Ann. Math.* **189** (2019), 527–604.

[233] Z. Gao and B. Klingler, The Ax–Schanuel conjecture for variations of mixed Hodge structures, arXiv:2101.10938.

[234] C. Gasbarri, Rational vs transcendental points on analytic Riemann surfaces, arXiv:1806.10844.

[235] É. Gaudron and G. Rémond, Théorème des périodes et degrés minimaux d'isogénies, *Comment. Math. Helv.* **89** (2014), 343–403.

[236] G. van der Geer, Siegel modular forms and their applications, *The 1-2-3 of modular forms,* J. H. Brunier, G. van der Geer, G. Harder, and D. Zagier, eds., Springer, Berlin, 2008, 181–245.

[237] M. Gevrey, Sur la nature analytique des solutions des équations aux dérivées partielles: Premier mémoire, *Ann. Sci. Ec. Norm. Super.* **35** (1918), 129–190.

[238] D. Ghioca, F. Hu, T. Scanlon, and U. Zannier, A variant of the Mordell–Lang conjecture, *Math. Res. Lett.* **26** (2019), 1383–1392.

[239] D. Ghioca, H. Krieger, and K. Nguyen, A case of the dynamical André–Oort conjecture, *Int. Math. Res. Not.* **2016**, 738–758.

[240] D. Ghioca, D. Masser, and U. Zannier, Bounded height conjecture for function fields, *New York J. Math.* **21** (2015), 837–846.

[241] D. Ghioca and K. D. Nguyen, A dynamical variant of the Pink-Zilber conjecture, *Algebra Number Theory* **12** (2018), 1749–1771.

[242] M. Giacomini, Holomorphic curves in Shimura varieties, *Arch. Math.* **111** (2018), 379–388. Correction, *Arch. Math.* **114** (2020), 119–121.

[243] M. Giacomini, Cyclotomic torsion points in elliptic schemes, arXiv:1802.02386.

[244] M. Giacomini, Holomorphic curves in mixed Shimura varieties, arXiv: 1910.14484.

[245] G. Glebov, On Chudnovsky-Ramanujan type formulae, M.Sc. Thesis, Simon Fraser University, 2016.

[246] D. Goldfeld, Gauss' class number problem for imaginary quadratic fields, *Bull. Am. Math. Soc.* **13** (1985), 23–37.

[247] R. L. Graham, B. L. Rothschild, and J. H. Spencer, *Ramsey theory*, 2nd ed., Wiley, New York, 1990.

[248] M. Gromov, Entropy, homology and semi-algebraic geometry [after Y. Yomdin], Séminaire Bourbaki, 1985–86, exposé 663, *Astérisque* **145–146** (1987), 225–240.

[249] A. Grothendieck, Esquisse d'un programme, *Geometric Galois actions, 1: Around Grothendieck's Esquisse d'un programme*, reprinted 5–48, with an English translation 243–283, L. Schneps and P. Lochak, eds., LMS Lecture Notes **242**, Cambridge University Press, Cambridge, 1997.

[250] J. Gwoździewicz, K. Kurdyka, and A. Parusiński, On the number of solutions of an algebraic equation on the curve $y = e^x + \sin x, x > 0$, and a consequence for o-minimal structures, *Proc. Am. Math. Soc.* **127** (1999), 1057–1064.

[251] P. Habegger, Intersecting subvarieties of $\mathbf{G}_m^n$ with algebraic subgroups, *Math. Ann.* **342** (2008), 449–466.

[252] P. Habegger, On the bounded height conjecture, *Int. Math. Res. Not.* **2009**, 860–886.

[253] P. Habegger, Intersecting subvarieties of abelian varieties with algebraic subgroups of complementary dimension, *Invent. Math.* **176** (2009), 405–447.

[254] P. Habegger, Weakly bounded height on modular curves, *Acta Math. Vietnam.* **35** (2010), 43–69.

[255] P. Habegger, Torsion points on elliptic curves in Weierstrass form, *Ann. Sc. Norm. Super. Pisa Cl. Sci. (5)* **12** (2013), 687–715.

[256] P. Habegger, Special points on fibred powers of elliptic surfaces, *J. Reine Angew. Math.* **685** (2013), 143–179.

[257] P. Habegger, Effective height upper bounds on algebraic tori, *Autour de la conjecture de Zilber–Pink*, Course notes, CIRM, 2011, 167–242, Panoramas et Synthèses **52**, P. Habegger, G. Rémond, T. Scanlon, E. Ullmo, and A. Yafaev, eds., Société Mathématique de France, Paris, 2017.

[258] P. Habegger, The Manin–Mumford conjecture, an elliptic curve, its torsion points and their Galois orbits, *O-minimality and diophantine geometry,* 1–40, G. O. Jones and A. J. Wilkie, eds., LMS Lecture Notes Series **421**, Cambridge University Press, Cambridge, 2015.

[259] P. Habegger, Singular moduli that are algebraic units, *Algebra Number Theory* **9** (2015), 1515–1524.

[260] P. Habegger, Diophantine approximation on definable sets, *Sel. Math. New Ser.* **24** (2018), 1633–1675.

[261] P. Habegger, The norm of Gaussian periods, *Q. J. Math.* **69** (2018), 153–182.

[262] P. Habegger, G. Jones, and D. Masser, Six unlikely intersection problems in search of effectivity, *Math. Proc. Cambridge Philos Soc.* **162** (2017), 447–477.

[263] P. Habegger and J. Pila, Some unlikely intersections beyond André–Oort, *Compos. Math.* **148** (2012), 1–27.

[264] P. Habegger and J. Pila, O-minimality and certain atypical intersection, *Ann. Sci. Ec. Norm. Super.* **49** (2016), 813–858.

[265] G. H. Hardy, *Orders of infinity,* Cambridge Tracts in Mathematics and Mathematical Physics **12**, Cambridge University Press, Cambridge, 1910. Reprinted by Hafner, New York, 1971.

[266] G. H. Hardy and E. M. Wright, *An introduction to the theory of numbers,* 5th ed., Oxford University Press, Oxford, 1979.

[267] D. Haskell and D. Macpherson, A version of o-minimality for the *p*-adics, *J. Symb. Log.* **62** (1997), 1075–1092.

[268] D. R. Hast, Functional transcendence for the unipotent Albanese map, arXiv:1911.00587.

[269] D. R. Heath-Brown, Cubic forms in ten variables, *Proc. London Math. Soc.* **47** (1983), 225–257.

[270] D. R. Heath-Brown, Counting rational points on cubic surfaces, *Astérisque* **251** (1998), 13–29.

[271] D. R. Heath-Brown, The density of rational points on curves and surfaces, *Ann. Math.* **155** (2002), 553–595.

[272] D. R. Heath-Brown, Counting rational points on algebraic varieties, *Analytic number theory, Cetraro, 2002,* 51–95, Lecture Notes in Mathematics **1891**, A. Perelli and C. Viola, eds., Springer, Berlin, 2006.

[273] H. Helfgott and A. Venkatesh, Integral points on elliptic curves and 3-torsion in class groups, *J. Am. Math. Soc.* **19** (2006), 527–550.

[274] C. W. Henson and L. Rubel, Some applications of Nevanlinna theory to mathematical logic: Identities of exponential functions, *Trans. Am. Math. Soc.* **282** (1984), 1–32.

[275] G. Hermann, Die Frage der endlich vielen Schritte in der Theorie der Polynomideale, *Math. Ann.* **95** (1926), 736–788.

[276] S. Herrero, R. Menares, and J. Rivera-Letelier, There are at most finitely many singular moduli that are *S*-units, arXiv:2102.05041.

[277] P. Hieronymi and C. Miller, Metric dimensions and tameness in expansions of the real field, *Trans. Am. Math. Soc.* **373** (2020), 849–874.

[278] M. Hindry, Autour d'une conjecture de Serge Lang, *Invent. Math.* **94** (1988), 575–603.

[279] M. Hindry, Introduction to abelian varieties and the Mordell–Lang conjecture, *Model theory and algebraic geometry,* 85–100, Lecture Notes in Mathematics **1696**, E. Bouscaren, ed., Springer, Berlin, 1998.

[280] M. Hindry and J. H. Silverman, *Diophantine geometry: An introduction,* Graduate Texts in Mathematics **201**, Springer-Verlag, New York, 2000.

[281] C. Hooley, On another sieve method and the numbers that are a sum of two h-th powers, *Proc. London Math. Soc.* **43** (1981), 73–109.

[282] E. Hrushovski, The Mordell–Lang conjecture for function fields, *J. Am. Math. Soc.* **9** (1996), 667–690.

[283] E. Hrushovski, The Manin–Mumford conjecture and the model theory of difference fields, *Ann. Pure Appl. Logic* **112** (2001), 43–115.

[284] E. Hrushovski and Y. Peterzil, A question of van den Dries and a theorem of Lipshitz and Robinson; not everything is standard, *J. Symb. Log.* **72** (2007), 119–122.

[285] P. Hubschmid, The André–Oort conjecture for Drinfeld modular varieties, *Comp. Math.* **149** (2013), 507–567.

[286] P. Hubschmid and E. Viada, An addendum to the elliptic torsion anomalous conjecture in codimension 2, *Rend. Semin. Mat. Univ. Padova* **141** (2019), 209–220.

[287] D. Husemöller, *Elliptic curves,* Graduate Texts in Mathematics **111**, Springer, New York, 1987.

[288] M. N. Huxley, The integer points in a plane curve, *Funct. Approx. Comment. Math.* **37** (2007), 213–231.

[289] D. T. Huynh, R. Sun, and S.-Y. Xie, Big Picard theorem for jet differentials and non-archimedean Ax–Lindemann theorem, arXiv:2012.12656.

[290] J.-M. Hwang and W.-K. To, Volumes of complex analytic subvarieties of Hermitian symmetric spaces, *Am. J. Math.* **124** (2002), 1221–1246.

[291] V. Jarnik, Über die Gitterpunkte auf konvexen Curven, *Math. Z.* **24** (1926), 500–518.

[292] B.-G. Jeon, On the number of hyperbolic Dehn fillings of a given volume, *Trans. Am. Math. Soc.* **374** (2021), 3947–3969.

[293] G. O. Jones, Improving the bound in the Pila-Wilkie theorem for curves, *O-minimality and diophantine geometry,* 204–215, G. O. Jones and A. J. Wilkie, eds., LMS Lecture Notes Series **421**, Cambridge University Press, Cambridge, 2015.

[294] G. O. Jones, D. J. Miller, and M. E. M. Thomas, Mildness and the density of rational points on certain transcendental curves, *Notre Dame J. Formal Logic* **52** (2011), 67–74.

[295] G. O. Jones and H. Schmidt, Pfaffian definitions of Weierstrass elliptic functions, *Math. Ann.* **379** (2021), 825–864.

[296] G. O. Jones and H. Schmidt, A Manin–Mumford theorem for the maximal compact subgroup of a universal vectorial extension of a product of elliptic curves, *Int. Math. Res. Not.* **2021**, 16203–16228.

[297] G. O. Jones and P. Speissegger, Generating the Pfaffian closure with total Pfaffian functions, *J. Log. Anal.* **4** (2012), Paper 5, 6pp.

[298] G. O. Jones and M. E. M. Thomas, The density of algebraic points on certain Pfaffian surfaces, *Q. J. Math.* **63** (2012), 637–651.

[299] G. O. Jones and M. E. M. Thomas, Rational values of Weierstrass zeta functions, *Proc. Edinburgh Math. Soc.* **59** (2016), 945–958.

[300] G. O. Jones and M. E. M. Thomas, Effective Pila-Wilkie bounds for unrestricted Pfaffian surfaces, *Math. Ann.* **381** (2021), 729–767.

[301] T. Kaiser, The Riemann mapping theorem for semi-analytic domains and o-minimality, *Proc. London Math. Soc.* **98** (2009), 427–444.

[302] E. Katz, J. Rabinoff, and D. Zureick-Brown, Uniform bounds for the number of rational points on curves of small Mordell-Weil rank, *Duke Math. J.* **165** (2016), 3189–3240.

[303] A. G. Khovanskii, *Fewnomials,* Translations of Math. Monographs **88**, AMS, Providence, RI, 1991.

[304] J. Kirby, The theory of the exponential differential equations of semi-abelian varieties, *Sel. Math. New Ser.* **5** (2009), 445–486.

[305] J. Kirby, Finitely presented exponential fields, *Algebra Number Theory* **7** (2013), 943–980.

[306] J. Kirby and B. Zilber, The uniform Schanuel conjecture over the real numbers, *Bull. London Math. Soc.* **38** (2006), 568–570.

[307] J. Kirby and B. Zilber, Exponentially closed fields and the conjecture on intersections with tori, *Ann. Pure Appl. Logic* **165** (2014), 1149–1168.

[308] B. Klingler, Hodge loci and atypical intersections: Conjectures, *Motives and complex multiplication,* Proceedings of a summer school at ETH, Zurich, J. Fresán, D. Jetchev, P. Jossen, and R. Pink, eds., Progress in Mathematics, Birkhauser, to appear.

[309] B. Klingler, E. Ullmo, and A. Yafaev, The hyperbolic Ax–Lindemann-Weierstrass conjecture, *Publ. Math. Inst. Hautes Etudes Sci.* **123** (2016), 333–360.

[310] B. Klingler, E. Ullmo, and A. Yafaev, Bi-algebraic geometry and the André–Oort conjecture, *Proceedings of the 2015 AMS Summer Institute in Algebraic Geometry,* 319–359, Proc. Symp. Pure Math. **97-2**, T. de Fernex, ed., AMS, Providence, RI, 2018.

[311] B. Klingler and A. Yafaev, The André–Oort conjecture, *Ann. Math.* **180** (2014), 867–925.

[312] J. Knight, A. Pillay, and C. Steinhorn, Definable sets in ordered structures. II, *Trans. Am. Math. Soc.* **295** (1986), 593–605.

[313] B. Kocel-Cynk, W. Pawłucki, and A. Valette, C^p-parametrization in o-minimal structures, *Can. Math. Bull.* **62** (2019), 99–108.

[314] M. Kontsevich and D. Zagier, Periods, *Mathematics unlimited – 2001 and beyond,* 771–808, B. Engquist and W. Schmid, eds., Springer, Berlin, 2001.

[315] P. Kowalski, Ax–Schanuel conditions in arbitrary characteristic, *J. Inst. Math. Jussieu* **18** (2019), 1157–1213.

[316] S. G. Krantz and H. R. Parks, *A primer of real analytic functions,* 2nd ed., Birkhauser, Boston, MA, 2002.

[317] L. Kühne, An effective result of André–Oort type, *Ann. Math.* **176** (2012), 651–671.

[318] L. Kühne, An effective result of André–Oort type II, *Acta Arith.* **161** (2013), 1–19.

[319] L. Kühne, The bounded height conjecture for semi-abelian varieties, *Compos. Math.* **156** (2020), 1405–1456.

[320] L. Kühne, Intersections of class fields, *Acta Arith.* **198** (2021), 109–127.

[321] L. Kühne, Equidistribution in families of abelian varieties and uniformity, arXiv:2101.10272.

[322] K.-W. Lan, An example-based introduction to Shimura varieties, *Motives and complex multiplication,* Proceedings of a summer school at ETH, Zurich, J. Fresán, D. Jetchev, P. Jossen, and R. Pink, eds., Progress in Mathematics, Birkhauser, to appear.

[323] E. Landau, Bemerkung zum Heilbronnschen Satz, *Acta Arith.* **1** (1935), 2–18.

[324] S. Lang, Division points on curves, *Ann. Mat. Pura Appl. (4)* **70** (1965), 229–234.

[325] S. Lang, *Introduction to transcendental numbers,* Addison-Wesley, Reading MA, 1966.

[326] S. Lang, *Elliptic functions,* 2nd ed., Graduate Texts in Mathematics **112**, Springer, New York, 1987.

[327] S. Lang, *Fundamentals of diophantine geometry,* Springer-Verlag, New York, 1983.

[328] S. Lang, *Number theory III,* Diophantine geometry, Encyclopaedia of Mathematical Sciences **60**, Springer-Verlag, Berlin, Heidelberg, New York, 1991.

[329] M. Laurent, Équations diophantiennes exponentielles, *Invent. Math.* **78** (1984), 299–327.

[330] M. Laurent, Sur quelques résultats récent de transcendance, *Astérisque* **198-200** (1991), 209–230.

[331] B. Lawrence and W. Sawin, The Shafarevich conjecture for hypersurfaces in abelian varieties, arXiv:2004.09046.

[332] B. Lawrence and A. Venkatesh, Diophantine problems via p-adic period mappings, *Invent. Math.* **221** (2020), 893–999.

[333] O. Le Gal and J.-P. Rolin, An o-minimal structure which does not admit C^∞ cellular decomposition, *Ann. Inst. Fourier* **59** (2009), 543–562.

[334] Y. Li, Singular units and isogenies between CM elliptic curves, arXiv:1810.13214.

[335] Q. Lin and M.-X. Wang, Isogeny orbits in a family of abelian varieties, *Acta Arith.* **170** (2015), 161–173.

[336] J.-M. Lion, C. Miller, and P. Speissegger, Differential equations over polynomially bounded bo-minimal structures, *Proc. Am. Math. Soc.* **131** (2002), 175–183.

[337] S. Łojasiewicz, *Introduction to complex analytic geometry,* Birkhauser, Basel, 1991.

[338] D. Lombardo, On the analytic bijections of the rationals in $[0, 1]$, *Rend. Lincei-Mat. Appl.* **28** (2017), 65–83.

[339] F. Luca and A. Riffault, Linear independence of powers of singular moduli, *Bull. Aust. Math. Soc.* **99** (2019), 42–50.

[340] A. Macintyre and A. Wilkie, On the decidability of the real exponential field, *Kreiseliana,* 441–467, A K Peters, P. Odifreddi, ed., Wellesley, MA, 1996.

[341] K. Mahler, On algebraic differential equations satisfied by automorphic functions, *J. Aust. Math. Soc.* **10** (1969), 445–450.

[342] H. B. Mann, On linear relations between roots of unity, *Mathematika* **12** (1965), 107–117.

[343] V. Mantova, Polynomial-exponential equations and Zilber's conjecture, with an appendix by Mantova and U. Zannier, *Bull. London Math. Soc.* **48** (2016), 309–320.

[344] D. Marker, Khovanskii's theorem, *Algebraic model theory (Toronto, ON, 1996)*, 181–193, Nato Adv. Sci. Inst. Ser. C Math. Phys. Sci. **496**, Bradd T. Hart, Alistair H. Lachlan and Matthew A. Valeriote, eds., Kluwer Academic, Dordrecht, 1997.

[345] D. Marker, *Model theory: An introduction,* Graduate Texts in Mathematics **217**, Springer-Verlag, New York, 2002.

[346] D. Marker, A remark on Zilber's exponentiation, *J. Symb. Log.* **71** (2006), 791–798.

[347] O. Marmon, A generalization of the Bombieri-Pila determinant method, *J. Math. Sci.* **171** (2010), 736–744.

[348] C. Martínez, The number of maximal torsion cosets in subvarieties of tori, *J. Reine Angew. Math.* **755** (2019), 103–126.

[349] D. Masser, Specialization of endomorphism rings of abelian varieties, *Bull. Soc. Math. Fr.* **124** (1996), 457–476.

[350] D. Masser, Heights, transcendence, and linear independence on commutative group varieties, *Lecture notes in Mathematics* **1819**, 1–51, F. Amoroso and U. Zannier, eds., Springer-Verlag, Berlin, 2003.

[351] D. Masser, Rational values of the Riemann zeta function, *J. Number Theory* **131** (2011), 2037–2046.

[352] D. Masser, Uniformity in unlikely intersections: An example for lines in three dimensions, Appendix B in [544].

[353] D. Masser, Counting rational points on analytic curves: A transcendence approach, Appendix F in [544].

[354] D. Masser, *Auxiliary polynomials in number theory,* Cambridge Tracts in Mathematics **207**, Cambridge University Press, Cambridge, 2016.

[355] D. Masser and G. Wüstholz, Isogeny estimates for abelian varieties, and finiteness theorems, *Ann. Math.* **137** (1993), 459–472.

[356] D. Masser and G. Wüstholz, Factorization estimates for abelian varieties, *Publ. Math. Inst. Hautes Etudes Sci.* **81** (1995), 5–24.

[357] D. Masser and U. Zannier, Torsion anomalous points and families of elliptic curves, *C. R. Acad. Sci. Paris, Ser. I* **346** (2008), 491–494, and *Am. J. Math.* **132** (2010), 1677–1691.

[358] D. Masser and U. Zannier, Torsion points on families of squares of elliptic curves, *Math. Ann.* **352** (2012), 453–484.

[359] D. Masser and U. Zannier, Bicycloyomic polynomials and impossible intersections, *J. Théor. Nr. Bordx.* **25** (2013), 635–659.

[360] D. Masser and U. Zannier, Torsion points on families of products of elliptic curves, *Adv. Math.* **259** (2014), 116–133.

[361] D. Masser and U. Zannier, Torsion points on families of simple abelian surfaces and Pell's equation over polynomial rings (with an appendix by E. V. Flynn), *J. Eur. Math. Soc.* **17** (2015), 2379–2416.

[362] D. Masser and U. Zannier, Abelian varieties isogenous to no Jacobian, *Ann. Math.* **191** (2020), 635–674.

[363] D. Masser and U. Zannier, Torsion points, Pell's equations, and integration in elementary terms, *Acta Math.* **225** (2020), 1157–1213.

[364] G. Maurin, Courbes algébriques et équations multiplicatives, *Math. Ann.* **341** (2008), 789-824.

[365] G. Maurin, Équations multiplicatives sur les sous-variétés des tores, *Int. Math. Res. Not.* **2011**, 5259–5366.

[366] N. Mavraki, Impossible intersections in a Weierstrass family of elliptic curves, *J. Number Theory* **169** (2016), 21–40.

[367] B. Mazur, Arithmetic on curves, *Bull. Am. Math. Soc.* **14** (1986), 207–259.

[368] B. Mazur, Questions about isogenies, automorphisms and bounds, 20 July 2015, available from the author's webpage.

[369] M. McQuillan, Division points on semi-abelian varieties, *Invent. Math.* **120** (1995), 143–159.

[370] C. Miller, Exponentiation is hard to avoid, *Proc. Am. Math. Soc.* **122** (1994), 257–259.

[371] C. Miller, A growth dichotomy for o-minimal expansions of ordered fields, *Logic: From foundations to applications,* 385–399, Proceedings of Logic Colloquium 1993, W. Hodges, M. Hyland, C. Steinhorn, and J. Truss, eds., Oxford University Press, New York, 1996.

[372] J. S. Milne, Abelian varieties, *Arithmetic geometry,* 130–150 G. Cornell and J. H. Silverman, eds., Springer, New York, 1986.

[373] J. S. Milne, Canonical models of (mixed) Shimura varieties and automorphic vector bundles, *Automorphic forms, Shimura varieties, and L-functions, volume 1 (Ann Arbor, MI, 1988),* Perspect. Math. **10**, L. Clozel and J.S. Milne, eds., Academic Press, Boston, MA, 1990, 283–414. Corrected version available on J. S. Milne's webpage.

[374] J. S. Milne, Introduction to Shimura varieties, *Harmonic analysis, the trace formula, and Shimura varieties,* 265–378, Clay Mathematics Proceedings **4**, J. Arthur, D. Ellwood and R Kottwitz, eds., AMS, Providence, RI, 2005. Corrected version (2017) available from the author's web page.

[375] J. S. Milne, What is … a Shimura variety? *Not. Am. Math. Soc.* **59** (2012), 1560–1561.

[376] N. Mok, Zariski closures of images of algebraic subsets under the uniformization map on finite-volume quotients of the complex ball, *Compos. Math.* **155** (2019), 2129–2149.

[377] N. Mok, J. Pila, and J. Tsimerman, Ax–Schanuel for Shimura varieties, *Ann. Math.* **189** (2019), 945–978.

[378] B. J. J. Moonen, Linearity properties of Shimura varieties, I, *J. Algebraic Geom.* **7** (1998), 539–567.

[379] B. J. J. Moonen, Linearity properties of Shimura varieties, II, *Compos. Math.* **114** (1998), 3–35.

[380] B. J. J. Moonen and F. Oort, The Torelli locus and special subvarieties, *Handbook of moduli, volume 2,* 549–594, Adv. Lect. Math. **25**, G. Farkas and I. Morrison, eds., International Press, Somerville, MA, 2013.

[381] L. J. Mordell, On the rational solutions of the indeterminate equation of third and fourth degress, *Math. Proc. Cambridge Philos Soc.* **21** (1922), 179–192.

[382] D. Mumford, A note on Shimura's paper "Discontinuous groups and abelian varieties", *Math. Ann.* **181** (1969), 345–351.

[383] D. Mumford, *Algebraic geometry I: Complex projective varieties,* Grund. Math. Wiss. **221**, Springer, Berlin, 1976.

[384] D. Mumford, *The Tata lectures on theta, volume 1,* Progress in Mathematics **28**, Birkhauser, Boston, MA, 1983.

[385] Y. Nakkajima and Y. Taguchi, A generalization of the Chowla-Selberg formula, *J. Reine Angew. Math.* **419** (1991), 119–124.

[386] F. Najman, Isogenies of non-CM elliptic curves with rational *j*-invariants over number fields, *Math. Proc. Cambridge Philos Soc.* **164** (2018), 179–184.

[387] J. Nekovář and N. Schappacher, On the asymptotic behaviour of Heegner points, *Turk J. Math.* **23** (1999), 549–556.

[388] Y. V. Nesterenko, *Algebraic independence,* TIFR–Narosa, Mumbai, 2009.

[389] J. Noguchi, An application of value distribution theory for semi-abelian varieties to problems of Ax–Lindemann and Manin–Mumford types, *Accad. Naz. Lincei Rend. Lincei-Mat. Appl.* **29** (2018), 401–411.

[390] R. Noot, On Mumford's families of abelian varieties, *J. Pure Appl. Algebra* **157** (2001), 87–106.

[391] R. Noot, Correspondances de Hecke, action de Galois et la conjecture d'André–Oort, Séminaire Bourbaki (2004–2005), *Asterisque* **307** (2006), 165–197.

[392] F. Oort, "The" general case of S. Lang's conjecture, *Diophantine approximation and abelian varieties,* 117-122 S. J. Edixhoven and J.-H. Evertse, eds., Lecture Notes in Mathematics **1566**, Springer, Berlin, 1993.

[393] F. Oort, Some questions in algebraic geometry, June 1995 manuscript, available from the author's webpage, and *Open problems in arithmetic algebraic geometry,* Appendix 1, 263–283, Adv. Lect. Math. **46**, F. Oort, ed., International Press, Somerville, MA, 2019.

[394] F. Oort, Canonical lifts and dense sets of CM points, *Arithmetic geometry, Cortona, 1994,* 228–234, F. Catanese, ed., Symposia. Math., XXXVII, Cambridge University Press, Cambridge, 1997.

[395] F. Oort, Special points in Shimura varieties, an introduction, notes for Intercity Seminar, 14 November 2003, available from the author's webpage.

[396] F. Oort and J. Tsimerman, The André–Oort conjecture, *Open problems in arithmetic algebraic geometry,* 61–70, Adv. Lect. Math. **46**, F. Oort, ed., International Press, Somerville, MA, 2019.

[397] M. Orr, Families of abelian varieties with many isogenous fibres, *J. Reine Angew. Math.* **705** (2015), 211–231.

[398] M. Orr, Introduction to abelian varieties and the Ax–Lindemann theorem, *O-minimality and diophantine geometry,* 100–128, G. O. Jones and A. J. Wilkie, eds., LMS Lecture Note Series **421**, Cambridge University Press, Cambridge, 2015.

[399] M. Orr, Height bounds and the Siegel property, *Algebra Number Theory* **12** (2018), 455–478.

[400] M. Orr, Unlikely intersections with Hecke translates of a special subvariety, *J. Eur. Math. Soc.* **23** (2021), 1–28.

[401] A. Oswal, A non-archimedean definable Chow theorem, arXiv:2009.06134.

[402] G. Papas, Ax–Schanuel for Lie algebras, arXiv:1905.04364.

[403] M. Paredes and R. Sasyk, Uniform bounds for the number of rational points on varieties over global fields, arXiv:12174.

[404] R. Paulin, An explicit André–Oort type result for $\mathbb{P}^1(\mathbb{C}) \times \mathbb{G}_m(\mathbb{C})$ based on logarithmic forms, *Publ. Math. Debr.* **88** (2016), 21–33.

[405] R. Paulin, An explicit André–Oort type result for $\mathbb{P}^1(\mathbb{C}) \times \mathbb{G}_m(\mathbb{C})$, *Math. Proc. Cambridge Philos. Soc.* **159** (2015), 153–163.

[406] M. Paun and N. Sibony, Value distribution theory for parabolic Riemann surfaces, arXiv:1403.6596v5.

[407] W. Pawłucki, Review of "Sur certains sous-ensembles de éspace euclidien" by J.-Y. Charbonnel, MR1136600, Mathematical Reviews, AMS.

[408] F. Pazuki, Theta height and Faltings height, *Bull. Soc. Math. Fr.* **140** (2012), 19–49.

[409] Y. Peterzil and S. Starchenko, Expansions of algebraically closed fields in o-minimal structures, *Sel. Math. New Ser.* **7** (2001), 409–445.

[410] Y. Peterzil and S. Starchenko, Expansions of algebraically closed fields II, functions of several variables, *J. Math. Logic* **3** (2003), 1–35.

[411] Y. Peterzil and S. Starchenko, Uniform definability of the Weierstrass ℘ functions and generalized tori of dimension one, *Sel. Math. New Ser.* **10** (2004), 525–550.

[412] Y. Peterzil and S. Starchenko, Complex analytic geometry in a non-standard setting, *Model theory with applications to algebra and analysis*, 117–165, Z. Chatzidakis, D. Macpherson, A. Pillay, and A. J. Wilkie, eds., LMS Lecture Note Series **349**, Cambridge University Press, Cambridge, 2008.

[413] Y. Peterzil and S. Starchenko, Complex analytic geometry and analytic-geometric categories, *J. Reine Angew. Math.* **626** (2009), 39–74.

[414] Y. Peterzil and S. Starchenko, Tame complex geometry and o-minimality, *Proceedings ICM, Hyderabad, 2010, volume 2*, 58–81, R. Bhatia, A. Pal, G. Rangarajan, V. Srinivas and M. Vanninathan, eds., Hindustan Book Agency, New Delhi, 2010.

[415] Y. Peterzil and S. Starchenko, Definability of restricted theta functions and families of abelian varieties, *Duke Math. J.* **162** (2013), 731–765.

[416] Y. Peterzil and S. Starchenko, A note on o-minimal flows and the Ax-Lindemann-Weierstrass theorem for semi-abelian varieties over $\mathbb{C}$, *Enseign. Math.* **63** (2017), 251–261.

[417] Y. Peterzil and S. Starchenko, Algebraic and o-minimal flows on complex and real tori, *Adv. Math.* **333** (2018), 539–569.

[418] Y. Peterzil and S. Starchenko, O-minimal flows on nilmanifolds, *Duke Math. J.* **170** (2021), 3935–3976.

[419] J. Pila, Geometric postulation of a smooth curve and the number of rational points, *Duke Math. J.* **63** (1991), 449–463.

[420] J. Pila, Geometric and arithmetic postulation of the exponential function, *J. Aust. Math. Soc.* **54** (1993), 111–127.

[421] J. Pila, Integer points on the dilation of a subanalytic surface, *Q. J. Math.* **55** (2004), 207–223.

[422] J. Pila, Rational points on a subanalytic surface, *Ann. Inst. Fourier* **55** (2005), 1501–1516.

[423] J. Pila, The density of rational points on a Pfaff curve, *Ann. Fac. Sci. Toulouse Math.* **16** (2007), 635–645.

[424] J. Pila, On the algebraic points of a definable set, *Sel. Math. New Ser.* **15** (2009), 151–170.

[425] J. Pila, Rational points of definable sets and results of André-Oort – Manin-Mumford type, *Int. Math. Res. Not.* **2009**, 2476–2507.

[426] J. Pila, Counting rational points on a certain exponential-algebraic surface, *Ann. Inst. Fourier* **60** (2010), 489–514, and corrigendum *Ann. Inst. Fourier* **67** (2017), 1277–1278.

[427] J. Pila, O-minimality and the André-Oort conjecture for $\mathbb{C}^n$, *Ann. Math.* **173** (2011), 1779–1840.

[428] J. Pila, Modular Ax-Lindemann-Weierstrass with derivatives, *Notre Dame J. Formal Logic* **54** (2013; Oléron proceedings), 553–565.

[429] J. Pila, Special point problems with elliptic modular surfaces, *Mathematika* **60** (2014), 1–31.

[430] J. Pila, Functional transcendence via o-minimality, *O-minimality and diophantine geometry*, 66–99, G. O. Jones and A. J. Wilkie, eds., LMS Lecture Note Series **421**, Cambridge University Press, Cambridge, 2015.

[431] J. Pila, On a modular Fermat equation, *Comment. Math. Helv.* **92** (2017), 85–103.

[432] J. Pila and T. Scanlon, Effective transcendental Zilber-Pink, arXiv:2105.05845.

[433] J. Pila, A. N. Shankar, and J. Tsimerman, Canonical heights on Shimura varieties and the André-Oort conjecture, with an appendix, Frobenius structures and unipotent monodromy at infinity, by H. Esnault and M. Groechenig, arXiv:2109.08788.

[434] J. Pila and J. Tsimerman, The André-Oort conjecture for the moduli space of abelian surfaces, *Compos. Math.* **149** (2013), 204–214.

[435] J. Pila and J. Tsimerman, Ax-Lindemann for $\mathcal{A}_g$, *Ann. Math.* **179** (2014), 659–681.

[436] J. Pila and J. Tsimerman, Ax-Schanuel for the j-function, *Duke Math. J.* **165** (2016), 2587–2605.

[437] J. Pila and J. Tsimerman, Muliplicative relations among singular moduli, *Ann. Sc. Norm. Super. Pisa Cl. Sci. (4)* **17** (2017), 1357–1382.

[438] J. Pila and J. Tsimerman, Independence of CM points in elliptic curves, *J. Eur. Math. Soc.*, DOI 10.4171/JEMS/1161.

[439] J. Pila and A. J. Wilkie, The rational points of a definable set, *Duke Math. J.* **133** (2006), 591–616.

[440] J. Pila and U. Zannier, Rational points in periodic analytic sets and the Manin-Mumford conjecture, *Atti Accad. Naz. Lincei Rend. Lincei-Mat. Appl.* **19** (2008), 149–162.

[441] A. Pillay and C. Steinhorn, Definable sets in ordered structures, *Bull. Am. Math. Soc.* **11** (1984), 159–162.

[442] A. Pillay and C. Steinhorn, Definable sets in ordered structures I, *Trans. Am. Math. Soc.* **295** (1986), 565–592.

[443] A. Pillay and C. Steinhorn, Definable sets in ordered structures III, *Trans. Am. Math. Soc.* **309** (1988), 469–476.

[444] R. Pink, A combination of the conjectures of Mordell-Lang and André-Oort, *Geometric methods in algebra and number theory*, 251–282, F. Bogomolov and Y. Tschinkel, eds., Progress in Mathathematics **253**, Birkhauser, Boston, MA, 2005.

[445] R. Pink, A common generalization of the conjectures of André-Oort, Manin-Mumford, and Mordell-Lang, manuscript dated 17 April 2005 available from the author's webpage.

[446] B. Poizat, L'egalité au cube, *J. Symb. Logic.* **66** (2001), 1647–1676.

[447] G. Pólya, On the mean value theorem corresponding to a given linear homogeneous differential equation, *Trans. Am. Math. Soc.* **24** (1922), 312–324.

[448] A. van der Poorten, Transcendental entire functions mapping every number field into itself, *J. Aust. Math. Soc.* **8** (1968), 192–193.

[449] C. Qiu, The Manin-Mumford conjecture and the Tate-Voloch conjecture for a product of Siegel moduli spaces, arXiv:1903.02089.

[450] N. Ratazzi, Intersection de courbes et de sous-groupes et problèmes de minoration de dernière hauteur dans les variétés abéliennes C.M., *Ann. Inst. Fourier* **58** (2008), 1575–1633.

[451] M. Raynaud, Courbes sur une variété abélienne et points de torsion, *Invent. Math.* **71** (1983), 207–233.

[452] M. Raynaud, Sous-variétés d'une variété abélienne et points de torsion, *Arithmetic and geometry, volume 1,* 327–352, Progress in Mathematics **35**, M. Artin and J. Tate, eds., Birkhauser, Boston, MA, 1983.

[453] G. Rémond, Intersection de sous-groupes et de sous-variétés, I, *Math. Ann.* **333** (2005), 525–548.

[454] G. Rémond, Intersection de sous-groupes et de sous-variétés, II, *J. Inst. Math. Jussieu* **6** (2007), 317–348.

[455] G. Rémond, Intersection de sous-groupes et de sous-variétés, III, *Comment. Math. Helv.* **84** (2009), 835–863.

[456] G. Rémond and E. Viada, Problème de Mordell-Lang modulo certaines sous-variétés abéliennes, *Int. Math. Res. Not.* **2003**, 1915–1931.

[457] R. Richard, A two-dimensional arithmetic André-Oort problem, arXiv:1808.07900.

[458] R. Richard and A. Yafaev, Topological and equidistributional refinement of the André-Pink-Zannier conjecture at finitely many places, *C. R. Math.* **357** (2019), 231–235.

[459] R. Richard and A. Yafaev, Height functions on Hecke orbits and the generalised André-Pink-Zannier conjecture, arXiv:2109.13718.

[460] A. Riffault, Equations with powers of singular moduli, *Int. J. Number Theory* **15** (2019), 445–468.

[461] J. Robinson, Definability and decision problems in arithmetic, *J. Symb. Log.* **14** (1949), 98–114.

[462] J.-P. Rolin, P. Speissegger, and A. J. Wilkie, Quasianalytic Denjoy-Carleman classes and o-minimality, *J. Am. Math. Soc.* **16** (2003), 751–777.

[463] M. Rosen and J. H. Silverman, On the independence of Heegner points associated to distinct imaginary fields, *J. Number Theory* **127** (2007), 10–36.

[464] P. Salberger, On the density of rational and integral points on algebraic varieties, *J. Reine Angew. Math.* **606** (2007), 123–147.

[465] P. Sarnak, Torsion points on varieties and homology of abelian covers, manuscript, 1988.

[466] P. Sarnak and S. Adams, Betti numbers of congruence subgroups (with an appendix by Z. Rudnick), *Isr. J. Math.* **88** (1994), 31–72.

[467] T. Scanlon, Automatic uniformity, *Int. Math. Res. Not.* **2004**, 3317–3326.

[468] T. Scanlon, Local André-Oort conjecture for the universal abelian variety, *Invent. Math.* **163** (2006), 191–211.

[469] T. Scanlon, Counting special points: Logic, diophantine geometry and transcendence theory, *Current Events Bulletin*, AMS, 2011, also *Bull. Am. Math. Soc.* **49** (2012), 51–71.

[470] T. Scanlon, O-minimality as an approach to the André-Oort conjecture, *Autour de la conjecture de Zilber-Pink*, 111–165, Panoramas et Synthèses **52**, P. Habegger, G. Rémond, T. Scanlon, E. Ullmo, and A. Yafaev, eds., Société Mathématique de France, Paris, 2017.

[471] T. Scanlon, Algebraic differential equations from covering maps, *Adv. Math.* **330** (2018), 1071–1100.

[472] A. Schinzel, Reducibility of lacunary polynomials, X, *Acta Arith.* **53** (1989), 47–97.

[473] A. Schinzel, *Polynomials with special regard to reducibility,* Encyclopedia of Mathematics and its Applications, Cambridge University Press, Cambridge, 2000.

[474] H. Schmidt, Relative Manin-Mumford in additive extensions, *Trans. Am. Math. Soc.* **371** (2019), 6463–6486.

[475] H. Schmidt, Counting rational points and lower bounds for Galois orbits, *Atti Accad. Naz. Lincei Rend. Lincei-Mat. Appl.* **30** (2019), 497–509.

[476] H. Schmidt, A short note on Manin-Mumford, *Int. J. Number Theory*, to appear.

[477] W. M. Schmidt, Integer points on curves and surfaces, *Monatsh. Math.* **99** (1985), 45–72.

[478] T. Schneider, Arithmetische Unterschungen elliptischer Integrale, *Math. Ann.* **113** (1937), 1–13.

[479] H. A. Schwarz, Verallgemeinerung eines analytischen Fundamentalsatzes, *Ann. Mat. Pura Appl. (2)* **10** (1880), 129–136, reprinted in *Gesammelte mathematische Abhandlungen, volume 2*, 296–302, Springer, Berlin, 1890.

[480] A. Sedunova, On the Bombieri-Pila method over function fields, *Acta Arith.* **181** (2017), 321–331.

[481] A. Seidenberg, Abstract differential algebra and the analytic case, *Proc. Am. Math. Soc.* **9** (1958), 159–164.

[482] A. Seidenberg, Abstract differential algebra and the analytic case, II, *Proc. Am. Math. Soc.* **23** (1969), 689–691.

[483] A. Seidenberg, Constructions in algebra, *Trans. Am. Math. Soc.* **197** (1974), 273–313.

[484] J.-P. Serre, *A course in arithmetic,* Graduate Texts in Mathematics **7**, Springer, New York, 1973.

[485] J.-P. Serre, *Algebraic groups and class fields,* Graduate Texts in Mathematics **117**, Springer, New York, 1988.

[486] J.-P. Serre, *Lectures on the Mordell-Weil theorem,* Aspects of Mathematics **E15**, Vieweg, Wiesbaden, 1989.

[487] H. Shiga and J. Wolfart, Criteria for complex multiplication and transcendence properties of automorphic functions, *J. Reine Angew. Math.* **463** (1995), 1–25.

[488] G. Shimura, On analytic families of polarized abelian varieties and automorphic functions, *Ann. Math.* **78** (1963), 149–192.

[489] G. Shimura, Construction of class fields and zeta functions of algebraic curves, *Ann. Math.* **85** (1967), 58–159.

[490] A. Shkop, Henson and Rubel's theorem for Zilber's pseudoexponentiation, *J. Symb. Log.* **77** (2012), 423–432.

[491] C.-L. Siegel, Uber die Classenzahl quadratischer Zahlkorper, *Acta Arith.* **1** (1935), 83–86.

[492] P. Speissegger, The Pfaffian closure of an o-minimal structure, *J. Reine Angew. Math.* **508** (1999), 189–211.

[493] H. Spence, Ax-Lindemann and André-Oort for a non-holomorphic modular function, *J. Théor. Nr. Bordx.* **30** (2018), 743–779.

[494] H. Spence, A modular André-Oort statement with derivatives, *Proc. Edinburgh Math. Soc.* **62** (2019), 323–365.

[495] H. Spence, Effective André-Oort type results for almost holomorphic modular forms, arXiv:1904.03432.

[496] M. Stoll, Uniform bounds for the number of rational points on hyperelliptic curves of small Mordell-Weil rank, *J. Eur. Math. Soc.* **21** (2019), 923–956.

[497] M. Stoll, Simultaneous torsion in the Legendre family, *Exp. Math.* **26** (2017), 446–459.

[498] A. Surroca, Valeurs algébriques de fonctions transcendantes, *Int. Math. Res. Not.* **2006**, Art. ID 16834, 31pp.

[499] A. Tarski, *A decision method for elementary algebra and geometry,* 2nd ed., revised, Rand Corporation, Berkeley, 1951.

[500] M. E. M. Thomas, An o-minimal structure without mild parameterization, *Ann. Pure Appl. Logic* **162** (2011), 409–418.

[501] R. Tijdeman, On the number of zeros of general exponential polynomials, *Nederl. Akad. Wetensch. Proc. Ser. A* **74** = *Indag. Math* **33** (1971), 1–7.

[502] P. Tretkoff, *Complex ball quotients and line arrangements in the projective plane,* Mathematical Notes **51**, Princeton University Press, Princeton, 2016.

[503] J. Tsimerman, The existence of an abelian variety over $\overline{\mathbb{Q}}$ isogenous to no Jacobian, *Ann. Math.* **176** (2012), 637–650.

[504] J. Tsimerman, Brauer-Siegel for arithmetic tori and lower bounds for Galois orbits of special points, *J. Am. Math. Soc.* **25** (2012), 1091–1117.

[505] J. Tsimerman, Ax-Schanuel and o-minimality, *O-minimality and diophantine geometry,* 216–221, G. O. Jones and A. J. Wilkie, eds., LMS Lecture Note Series **421**, Cambridge University Press, Cambridge 2015.

[506] J. Tsimerman, The André-Oort conjecture for $\mathcal{A}_g$, *Ann. Math.* **187** (2018), 379–390.

[507] P. Tzermias, The Manin-Mumford conjecture: A brief survey, *Bull. London Math. Soc.* **32** (2000), 641–652.

[508] E. Ullmo, Manin-Mumford, André-Oort, the equidistribution point of view, Note de cours à l'école d'été, *Equidistribution in number theory, Montreal 2005,* 103–138, Nato Sci. Ser. II, Math. Phys. Chem. **237**, A. Granville and Z. Rudnick, eds., Springer, Dordrecht, 2007.

[509] E. Ullmo, Quelques applications du théorème de Ax-Lindemann hyperbolique, *Compos. Math.* **150** (2014), 175–190.

[510] E. Ullmo, Structures Spéciales et problème de Zilber-Pink, *Autour de la conjecture de Zilber-Pink,* 1–30, Course Notes, CIRM, 2011, *Panoramas et Synthèses* **52**, P. Habegger, G. Rémond, T. Scanlon, E. Ullmo, and A. Yafaev, eds., Société Mathématique de France, Paris, 2017.

[511] E. Ullmo and A. Yafaev, Galois orbits and equidistribution of special subvarieties: Towards the André-Oort conjecture, *Ann. Math.* **180** (2014), 823–865.

[512] E. Ullmo and A. Yafaev, A characterization of special subvarieties, *Mathematika* **57** (2011), 263–273.

[513] E. Ullmo and A. Yafaev, Nombre de classes des tores de multiplication complexe et bornes inférieures pour orbites Galoisiennes de points spéciaux, *Bull. Soc. Math. Fr.* **143** (2015), 197–228.

[514] E. Ullmo and A. Yafaev, Hyperbolic Ax-Lindemann in the cocompact case, *Duke Math. J.* **163** (2014), 433–463.

[515] E. Ullmo and A. Yafaev, Algebraic flows on abelian varieties, *J. Reine Angew. Math.* **741** (2018), 47–66.

[516] E. Ullmo and A. Yafaev, O-minimal flows on abelian varieties, *Q. J. Math.* **68** (2017), 359–367.

[517] E. Ullmo and A. Yafaev, Holomorphic curves in compact Shimura varieties, *Ann. Inst. Fourier* **68** (2018), 647–659.

[518] D. Urbanik, Effective methods for diophatine finiteness, arXiv:2110.14829.

[519] S. Van Hille, Smooth parameterizations of power-subanalytic sets and compositions of Gevrey functions, arXiv:1905.06408.

[520] S. Van Hille, On a family of mild functions, *Int. J. Number Theory* **17** (2021), 1379–1390.

[521] S. Van Hille, Mild parameterizations of power-subanalytic sets, arXiv:2105.04918.

[522] F. Vermeulen, Points of bounded height on curves and the dimension growth conjecture over $\mathbb{F}_q[t]$, arXiv:2003.10988.

[523] E. Viada, The intersection of a curve with algebraic subgroups in a product of elliptic curves, *Ann. Sc. Norm. Super. Pisa Cl. Sci. (5)* **2** (2003), 47–75.

[524] E. Viada, The intersection of a curve with a union of translated co-dimension two subgroups in a power of an elliptic curve, *Algebra Number Theory* **2** (2008), 249–298.

[525] E. Viada, An explicit Manin-Demjanenko theorem in elliptic curves, *Can. J. Math.* **70** (2018), 1173–1200.

[526] M. Walsh, Bounded rational points on curves, *Int. Math. Res. Not.* **2015**, 5644–5658.

[527] A. J. Wilkie, Model completeness results for expansions of the ordered field of real numbers by restricted Pfaffian functions and the exponential function, *J. Am. Math. Soc.* **9** (1996), 1051–1094.

[528] A. J. Wilkie, Schanuel's conjecture and the decidability of the real exponential field, *Algebraic model theory (Toronto, ON, 1996)*, Bradd T. Hart, Alistair H. Lachlan and Matthew A. Valeriote, eds., 223–230, Nato Adv. Sci. Inst. Ser. C, Math. Phys. Sci. **496**, Kluwer Academic, Dordrecht, 1997.

[529] A. J. Wilkie, A theorem of the complement and some new o-minimal structures, *Sel. Math. New Ser.* **5** (1999), 397–421.

[530] A. J. Wilkie, Diophantine properties of sets definable in an o-minimal structure, *J. Symb. Log.* **69** (2004), 851–861.

[531] A. J. Wilkie, Rational points on definable sets, *O-minimality and diophantine geometry*, 41–65, G. O. Jones and A. J. Wilkie, eds., LMS Lecture Note Series 421, Cambridge University Press, Cambridge, 2015.

[532] G. Wüstholz, A note on the conjectures of André-Oort and Pink, with an appendix by Lars Kühne, *Bull. Inst. Math. Acad. Sin. (N. S.)* **9** (2014), 735–779.

[533] A. Yafaev, Special points on products of two Shimura curves, *Manuscr. Math.* **104** (2001), 163–171.

[534] A. Yafaev, A conjecture of Yves André's, *Duke Math. J.* **132** (2006), 393–407.

[535] A. Yafaev, The André-Oort conjecture: A survey, *L-functions and Galois representations,* 381–406, D. Burns and K. Buzzard, eds., pp. 381–406, LMS Lecture Note Series **320**, Cambridge University Press, Cambridge, 2007.

[536] Y. Yomdin, Volume growth and entropy, *Isr. J. Math.* **57** (1987), 285–300.

[537] Y. Yomdin, C^k-resolution of semi-algebraic mappings: Addendum to "Volume growth and entropy", *Isr. J. Math.* **57** (1987), 301–317.

[538] Y. Yomdin, Analytic reparameterization of semi-algebraic sets, *J. Complexity* **24** (2008), 54–76.

[539] Y. Yomdin, Smooth parameterizations in dynamics, analysis, diophantine and computational geometry, *Jpn. J. Ind. Appl. Math.* **32** (2015), 411–435.

[540] X. Yuan and S.-W. Zhang, On the averaged Colmez conjecture, *Ann. Math.* **187** (2018), 533–638.

[541] D. Zagier, Elliptic modular forms and their applications, *The 1-2-3 of modular forms,* 1–103 J. H. Brunier, G. van der Geer, G. Harder, and D. Zagier, eds., Springer, Berlin, 2008.

[542] U. Zannier, Vanishing sums of roots of unity, *Rend. Sem. Mat. Univ. Politec. Torino* **53** (1995), 487– 495.

[543] U. Zannier, Appendix in [473], 517–539.

[544] U. Zannier, *Some problems of unlikely intersections in arithmetic and geometry,* with appendices by D. Masser, Annals of Mathematics Studies **181**, Princeton University Press, Princeton, 2012.

[545] U. Zannier, Elementary integration of differentials in families and conjectures of Pink, *Proceedings ICM Seoul 2014, volume 2,* 531–555, Kyung Moon Sa, Seoul, 2014.

[546] S.-W. Zhang, Positive line bundles on arithmetic surfaces, *J. Am. Math. Soc.* **8** (1995), 187–221.

[547] S.-W. Zhang, Small points and Arakelov theory, *Proceedings ICM 1998, Documenta Math, extra volume,* Deutsche Mathematiker Vereinigung, Berlin Extra Volume ICM 1998, 217–225.

[548] S.-W. Zhang, Equidistribution of CM points on quaternion Shimura varieties, *Int. Math. Res. Not.* **2005**, 3657–3689.

[549] B. Zilber, Fields with pseudo-exponentiation, *Proceedings of an international conference in memory of A. Malt'sev,* Novosibirsk, 1999, arXiv:math/0012023.

[550] B. Zilber, Exponential sums equations and the Schanuel conjecture, *J. London Math. Soc. (2)* **65** (2002), 27–44.

[551] B. Zilber, Pseudo-exponentiation on algebraically closed fields of characteristic zero, *Ann. Pure Appl. Logic* **132** (2005), 67–95.

[552] B. Zilber, Covers of the multiplicative group of an algebraically closed field of characteristic zero, *J. London Math. Soc.* **74** (2006), 41–58.

[553] B. Zilber, Model theory of special subvarieties and Schanuel-type conjectures, *Ann. Pure Appl. Logic* **167** (2016), 1000–1028.

[554] B. Zilber and C. Daw, Modular curves and their pseudo-analytic cover, arXiv:2107.11110.

[555] R. Richard and A. Yafaev, Generalised André-Pink-Zannier conjecture for Shimura varieties of abelian type, arXiv:2111.11216.

[556] G. Baldi, B. Klingler, and E. Ullmo, On the geometric Zilber-Pink theorem and the Lawrence-Venkatesh method, arXiv:2112.13040.

[557] R. F. Coleman, Torsion points on curves, *Galois representations and arithmetic algebraic geometry*, 235–247, Advanced Studies in Pure Mathematics **12**, Y. Ihara, ed., North-Holland, Amsterdam, 1987.

[558] G. Binyamini, D. Novikov, and B. Zack, Wilkie's conjecture for Pfaffian structures, arXiv:2202.05305

List of Notation

1-motive, 140

$\mathcal{A}' = (\mathcal{A}, \ldots)$: expansion of structure, 64

$\mathcal{A} \equiv \mathcal{B}$: elementary equivalence of structures, 64

$A^{\vee}$: dual abelian variety of A, 172

$\mathcal{A}_g$: Siegel modular variety, 48

ACF_0: theory of algebraically closed fields of characteristic zero, 65

$B(k, n, d)$: vanishing order of alternant, 84

$\mathbb{B}$: the Zilber field, 153

$\mathbb{C}^{\times}$: multiplicative group of non-zero complex numbers, 3

$\mathbb{C}_{\mathrm{alg}}$: the structure of the complex field, 69

$D_k(d)$: sum of degrees of k-variate monomials of degree $\leq d$, 84

$\Delta(x)$: complexity of CM point $x \in \mathcal{A}_g$, 51

$\Delta(\sigma)$: complexity of singular modulus, 30

$\Delta(E)$: discriminant of elliptic curve, 27

$\Delta(T)$: complexity of special subvariety of $Y(1)^n$, 35

$\delta(V)$: the defect of V, 157

$\Delta(z)$: discriminant function, 4

$\delta_S(V)$: the defect of V, 157

$\delta_Z(A)$: Zariski defect, 124

DLO: theory of dense linear order without endpoints, 66

$e(z)$: modified exponential $e(z) = \exp(2\pi i z)$ and also its cartesian powers, 3

$E_4(z)$: weight 4 Eisenstein series, 4

$\mathrm{End}(A)$: endomorphism ring of abelian variety, 50

$\mathrm{End}^0 A$: endomorphism algebra of abelian variety, 50

$\exists$: universal logical quantifier, 61

F: classical fundamental domain of the modular group, 28

$\forall$: existential logical quantifier, 61

G_{tor}: subgroup of torsion points of an abelian group G, 18

$\mathrm{Gal}(L/M)$: the Galois group of L over M, 53

$\gamma(f, d)$: Bezout bound for f, 11

$\Gamma(N)$: principle congruence subgroup of modular group of level N, 36

$\mathbb{G}_a$: additive group, 179

$\mathbb{G}_m$: the multiplicative group, 18

$\mathrm{GL}_2^+(\mathbb{Q})$: elements of $\mathrm{GL}_2(\mathbb{Q})$ with positive determinant, 29

$\mathrm{GL}_2^+(\mathbb{R})$: elements of $\mathrm{GL}_2(\mathbb{R})$ with positive determinant, 29

$h(\alpha)$: absolute logarithmic Weil height, 15

$H(\mathbb{R})^c$: connected component of the identity in $H(\mathbb{R})$, 49

$h(D)$: class number of quadratic order of discriminant D, 31

$H(r)$: multiplicative height; Definition 2.1, 9

$h(O_D)$: class number of quadratic order, 31

$H_k^{\mathrm{poly}}(\alpha)$: polynomial height of an algebraic number, 89

h_{Fal}: Faltings height, 53

$\mathbb{H}$: complex upper half-plane, 3

$\mathbb{H}_g$: Siegel upper half-space, 49

241

$\langle V \rangle$: smallest special subvariety containing V, 157

$\langle V \rangle_S$: smallest special variety containing V, 157

$V^{\langle h \rangle}$: Zariski closure of incidence variety, 207

V^{oa}: open anomalous set of V, 166

$\mathcal{W}(Y)$: collection of weakly special subvarieties of weakly special $Y \subset X$, 159

$\mathcal{W} = \mathcal{W}_X$: collection of weakly special subvarieties of X, 150, 157

$\langle W \rangle_\mathcal{M}$: closure of W in designated collection $\mathcal{M}$, 119

$X^{[k]}$: union of special subvarieties of X of *codimension* at least k, 160

$\mathcal{X}_g$: universal family of (principally polarized) abelian varieties of dimension g (with suitable level structure), 149

$Y(1)$: the modular curve, 28

$(Y(1)^n)^{[k], \text{nss}}$: union of not strongly special subvarieties of $Y(1)^n$ of codimension at least k, 181

$Y(2)$: modular curve with level 2 structure, 37

$\mathcal{Z}$: collection of algebraic subvarieties of an algebraic domain, 120

$Z(\mathbb{Q}, T)$: set of rational points of Z up to multiplicative height T; Definition 2.2, 9

$Z(K, T)$: K-rational points of Z up to multiplicative height T, 95

$Z(k, T)$: set of degree at most k points of Z up to multiplicative height T; Definition 2.12, 15

$Z[f, h]$: degree f points of Z up to logarithmic height h, 104

Z^t: transpose of matrix Z, 49

Z^{alg}: the algebraic part of a set Z, 14

$Z^{\text{semi}}(k, T)$: semi-rational points in Z of degree $\leq k$ and height $\leq T$, 92

Z^{trans}: the transcendental part of a set Z, 14

Index